Springer Proceedings in Physics 34

Springer Proceedings in Physics

Managing Editor: H. K. V. Lotsch

Volumes 1 – 29 are listed on the back inside cover

Amorphous and Crystalline Silicon Carbide

and Related Materials

Proceedings of the First International Conference
Washington DC, December 10 and 11, 1987

Editors: G. L. Harris and C. Y.-W. Yang

With 156 Figures

Springer-Verlag Berlin Heidelberg New York
London Paris Tokyo

Dr. Gary L. Harris

Material Science Research Center of Excellence, Howard University,
2300 6th Street, N.W., Washington, DC, 20059, USA

Dr. Cary Y.-W. Yang

Santa Clara University, Santa Clara, CA 95053, USA

ISBN-13: 978-3-642-93408-7 e-ISBN-13: 978-3-642-93406-3
DOI: 10.1007/978-3-642-93406-3

Library of Congress Cataloging-in-Publication Data. Amorphous and crystalline silicon carbide and related materials: proceedings of the first international conference, Washington DC, December 10 and 11, 1987 / editors, G. L. Harris and C. Y.-W. Yang. (Springer proceedings in physics; 34) "Papers presented at the First International Conference on Amorphous and Crystalline Silicon Carbide and Related Materials (ICACSC 87), which was held at Howard University" – Pref. Sponsored by the National Science Foundation et al. Includes index. 1. Silicon carbide – Congresses. I. Harris, Gary Lynn, 1953–. II. Yang, C. Y.–W. (Carry Y.–W.), 1948– III. International Conference on Amorphous and Crystalline Silicon Carbide and Related Materials (1st: 1987: Howard University) IV. National Science Foundation (U.S.) V. Series: Springer proceedings in physics; V. 34.
TP245.S5A68 1989 621.381'71–dc 19 88-36473

Softcover reprint of the hardcover 1st edition 1989

Preface

This volume contains written versions of the papers presented at the First International Conference on Amorphous and Crystalline Silicon Carbide and Related Materials (ICACSC 87), which was held at Howard University on December 10 and 11, 1987. The conference followed the International Electron Devices Meeting. For almost 20 years there had been no formal international conference on SiC. The purpose of the conference was to provide an international forum for the discussion and exchange of ideas regarding the current status of research aimed at developing SiC semiconductor devices and circuits. It was attended by over one hundred participants from six countries.

Of the 29 papers presented at the conference 25 are reproduced here in their entirety while the remaining four are represented only by abstracts. Several invited papers provide a detailed review of recent results in silicon carbide in the United States, Europe, and Japan. The first and second parts of the volume report on the epitaxial and bulk growth of both amorphous and crystalline silicon carbide. Parts III and IV contain reports on characterization, processing, and device applications of silicon carbide and related materials. Several new and exciting results are reported in this volume for the first time, including second harmonic generation in SiC, high temperature MOSFET and MESFET silicon carbide devices, bulk crystal growth of 6H silicon carbide, and amorphous SiC emitter heterojunctions and bipolar transistors.

Finally, on behalf of all the participants, we would like to thank our sponsors – the National Science Foundation; Howard University Material Science Research Center of Excellence; IEEE Electron Devices Group, Washington Section; Honeywell Incorporated; 3M Corporation; and Xerox Corporation – for their generous financial support. Thanks are also due to the members of the International Advisory Committee and the 1987 Organizing Committee.

Washington, DC
September 1988

G.L. Harris
C.Y.-W. Yang

Contents

Part I

Growth of Crystalline Silicon Carbide

Recent Developments in SiC (USA)

J.A. Powell and L.G. Matus

NASA Lewis Research Center, Cleveland, Ohio 44135, USA

Significant progress has been made recently in the USA in the crystal growth and semiconductor device technology of SiC. Advances in the growth of both β (3C) SiC and 6H SiC have been made. Advances in device technology include the development of a 650°C metal-oxide-semiconductor field-effect transistor (MOSFET), and improved processes for ion implantation and plasma-assisted etching. In addition to these technological advances, there has been a growing awareness of the potential benefits of SiC semiconductor devices. This paper will present a discussion of these topics plus an introduction to SiC properties and a historical overview of SiC research in the USA. All discussions will be limited to developments in crystalline SiC.

1. Introduction

The development of SiC as a semiconductor has been pursued sporadically in the USA for more than thirty years. The driving force behind these efforts generally has been aerospace applications involving high temperature and high radiation environments. The trend toward greater use of electronic devices and sensors in propulsion and power systems has increased this interest in the development of SiC. The main obstacle to the development of SiC has been the lack of a reproducible process for growing semiconductor quality crystals. To provide a basis for appreciating the crystal growth difficulty and the benefits of SiC, its crystal structure and properties will be reviewed first.

2. Crystal Structure and Properties

An unusual aspect of SiC is that it exhibits a form of one-dimensional polymorphism called polytypism [1,2]. The many polytypes of SiC differ from one another only in the stacking sequence of double layers of Si and C atoms. Each double layer consists of a plane of close-packed Si atoms over a plane of close-packed C atoms; one Si atom lies directly over each C atom in a double layer. Each successive double layer is stacked on the previous double layer in a close-packed arrangement that allows for only three possible relative positions for the double layers. These positions are normally labeled A, B, and C. Depending on the stacking sequence, various structures (i.e. cubic, hexagonal, or rhombohedral) are produced. The stacking direction is the c-axis in the hexagonal frame of reference.

The more common SiC polytypes are listed in Table I. Several ways of designating the various structures are shown in this table. The most common is the Ramsdell notation which is a number followed by a letter. The number is the number of double layers in a stacking repeat sequence and the letter designates the structure. Thus, we have 3C for cubic SiC. All of the other polytypes are known as α-SiC. The most common is 6H SiC. There really are only two polytypes of practical importance, β-SiC (the only cubic form) and 6H SiC. A final comment about SiC crystal structure is that SiC has the same diamond cubic lattice arrangement as Si and GaAs. So, its lattice constant is usually given in the cubic frame of reference, and the 6H SiC lattice constants are given for the hexagonal frame of reference.

Springer Proceedings in Physics, Vol. 34 **Amorphous and Crystalline Silicon Carbide**
Editors: G.L. Harris and C.Y.-W. Yang

Table I. Selected SiC Polytypes

	Ramsdell Notation	Stacking Sequence
(cubic or beta)	3C	ABCABC...
(alpha)	6H	ABCACBABCACB...
	15R	ABCBACABACBCACBAB...
	4H	ABACABAC...
	2H	ABAB...

Table II. Comparison of Semiconductors

Property	Si	GaAs	GaP	β-SiC (6H SiC)	Diamond
Bandgap (eV) at 300 K	1.1	1.4	2.3	2.2 (2.9)	5.5
Maximum operating temperature (°C)	300	460	925	873 (1240)	1100(?)
Melting point (°C)	1420	1238	1470	Sublimes > 1800	Phase change
Physical stability	Good	Fair	Fair	Excellent	Very good
Electron mobility R.T., $cm^2/V-s$	1400	8500	350	1000 (600)	2200
Hole mobility R.T., $cm^2/V-s$	600	400	100	40	1600
Breakdown voltage E_b, 10^6 V/cm	.3	.4	—	4	10
Thermal conductivity σ_T, W/cm-°C	1.5	.5	.8	5	20
Sat. elec. drift vel. v(sat), 10^7 cm/s	1	2	—	2.5	2.7
Dielectric const., K	11.8	12.8	11.1	9.7	5.5
Relative Z_J	1	7	—	1100	8100
Relative Z_K	1	.5	—	6	32

$Z_J \propto E_b^2 \, v^2(sat)$ $\qquad Z_K \propto \sigma_T [v(sat)/K]^{1/2}$

An appreciation of the potential of SiC can be gained by examining Table II, which is a comparison of its properties with diamond and GaP (two other contenders for high temperature semiconductor applications) and the two most common commercially available semiconductors, Si and GaAs. In the SiC column of the table, if a property of 6H SiC differs from β-SiC, it is listed within parentheses. The maximum operating temperature was calculated relative to that of Si by assuming a maximum for Si of 300°C and multiplying this temperature (expressed in K) by the ratio of bandgaps. The maximum for diamond is imposed by a phase change.

Silicon carbide does not melt at any reasonable pressure, but does sublime at temperatures greater than 1800°C. Below 1500°C, its physical stability is excellent and its stability in an oxidizing atmosphere gives it an edge over diamond. We believe that long term stability at high temperatures will be a problem with the more common III-V compounds, such as GaAs and GaP. Thus, SiC has a significant advantage where long-term reliable operation is a requirement.

The room temperature electron and hole mobility of SiC does not compare as favorably as the other properties but is acceptable for most applications. For high temperature power applications, the advantages of SiC and diamond are illustrated by their much higher values of breakdown voltage and thermal conductivity.

Another potential area of application for SiC (and diamond) is high frequency. Two Figures of Merit for semiconductors for high frequency applications are the Johnson (Z_J) [3] and Keyes (Z_K) [4]. Relative values (to that of Si) for these quantities are shown in Table II and illustrate the high potential of SiC and diamond.

3. Potential Benefits of SiC Semiconductor Devices

The previously mentioned trend toward more electronic devices and sensors in propulsion and power systems is inexorable. The need for higher performance and efficiency requires more control over the many elements of these systems. Hence, there is a growing need for sensors (with integrated electronics), power electronics for actuators, and distributed microcomputer chips. Unfortunately, in many applications, the temperature (and sometimes radiation) environment is approaching or exceeding the limits of commercially available semiconductors. As the temperature rises, the reliability of devices goes down. An example is the automobile engine which is being designed to run hotter and at the same time use more electronic devices for control and diagnostics. This has forced the placement of some electronics in cooler locations away from the engine which in turn makes the electronics susceptible to electrical noise pickup.

At NASA, we see an increasing need for electronics which are capable of sustained operation at temperatures in excess of 400°C. In advanced turbine engines, there is a need for control and monitoring electronics that are not constrained to temperature-controlled environments. The goal is increased flexibility in the use of electronics, with the benefit being increased performance and elimination of cooling (weight savings). In space power systems, there is a need for high temperature power conversion electronics. In this application, higher operating temperatures lead to much smaller radiators that are used to get rid of waste heat due to power conversion losses.

Component reliability is a key issue in all aerospace applications. Failure can lead to tragic consequences. Because of its ceramic character, we believe that the potential reliability of semiconductor devices and sensors fabricated from SiC will be much higher than that obtainable from any other semiconductor material. This reliability advantage should hold for all operating temperatures.

Other potentially important applications for SiC are in light-emitting diodes (LEDs)(using 6H SiC) and in high frequency components. The high frequency applications could ultimately have the greatest commercial impact of all SiC applications considered so far.

With these potential benefits in mind, let's now review SiC developments in the USA, past and present. During this review, keep in mind that the development of SiC has really been an international effort, with much cross-fertilization of ideas and processes.

4. Historical Overview of SiC Research in the USA

Prior to the mid 1950's, SiC was only available through the industrial Acheson process for making abrasive material [5]. Initial research concentrated on crystal structure and optical studies [6]. Several events spurred a major USA SiC effort in the late 1950's. First, in 1955, a laboratory sublimation process was developed by Lely in F.R.Germany for growing α-SiC crystals [7]. Second, the supersonic transport and nuclear space power programs were initiated with U.S. Air Force and NASA funding. These programs funded many SiC research groups over the subsequent 15 years. The proceedings of three international conferences, held in the USA in 1959, 1968, and 1973, documents this work fairly well [8-10]. Institutions and companies that took part in this work include the A.F. Cambridge Research Laboratories, Westinghouse, General Electric, Stanford Research Institute (now SRI International), Carborundum, Norton, Hughes Research Laboratories, the NASA Lewis Research Center, and the NASA Electronics Research Center. During this 15 year period, much device technology was developed and prototype diodes and field-effect transistors (FETs) were produced [11]. Substrates used in this work were mostly α-SiC crystals produced by the sublimation (Lely) process.

The following is a description of the sublimation process as it had been developed by the early 1970's. The basic elements of the process are shown in Fig. 1. A hollow cavity was formed inside a charge of polycrystalline SiC. The charge was heated to about 2500°C in a graphite tube furnace at which point the SiC sublimed and condensed on slightly cooler parts of the cavity. Growth took place on a thin porous graphite cylinder which formed the wall of the cavity. Nucleation of individual crystals was uncontrolled, and the resulting crystals were randomly-sized hexagonally-shaped α-SiC platelets [12]. The platelets exhibited a layered structure of various α polytypes with the stacking direction of these various polytypes being along the c-axis. Frequently, the transition region between two polytypes in a platelet was a region of random stacking of double layers, that is, a region of one-dimensional disorder. The predominant polytype (generally more than 75%) was 6H SiC, followed by 15R and 4H. Since the energy bandgap is different for the various polytypes, undesirable heterojunctions are produced in crystals with multiple polytypes and disordered regions.

Because of the difficulties with the sublimation process, solution growth methods were tried [13-15]. Although the results were not satisfactory, small β-SiC crystals with very good electrical properties were produced [13]. Despite some progress, the chief obstacle to SiC development was the lack of a reproducible crystal growth process.

Along with the lack of progress in crystal growth, military and NASA funding for SiC came to end in the early 1970's. As a result, except for fundamental optical investigations at Westinghouse [6], SiC research ceased in the USA for the remainder of the 1970's.

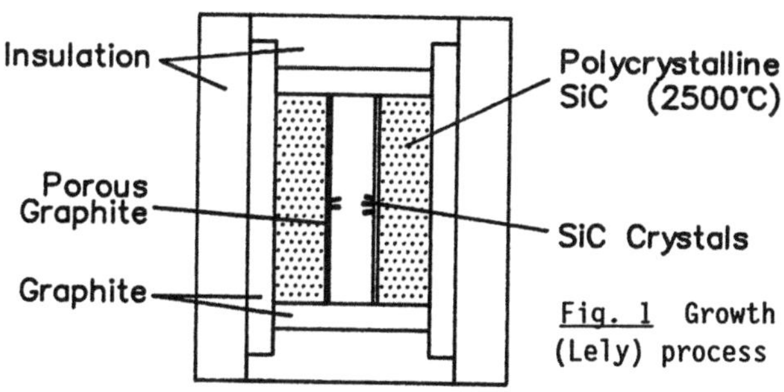

Fig. 1 Growth cavity used in the sublimation (Lely) process

5. Recent Progress in Crystal Growth

In 1980, NASA reinstated SiC research to answer the growing need for high temperature electronics. Shortly thereafter, the U.S. Navy also began sponsoring SiC research. Since then, developments in crystal growth have led to a resurgence of interest in SiC. Institutions currently active in SiC growth, characterization, and/or device development include the NASA Lewis Research Center, the Naval Research Laboratory (NRL), North Carolina State University (NCSU), Howard University, the University of Pittsburgh, Case Western Reserve University (CWRU), Hughes Research Laboratories, and Cree Research. Two growth processes have led the way. The first, in 1982, was a method for growing large-area β-SiC films on Si [16], and the second, in 1987, was the first successful implementation in the USA of a seeded-growth sublimation process [17]. Developments in these processes will now be described.

5.1 Heteroepitaxial Growth of SiC by CVD

Since large-area SiC crystals were not available, researchers had long tried to grow SiC on other substrates, in particular, on Si. These efforts were largely unsuccessful because of a large lattice mismatch (e.g. the SiC lattice is 20% smaller than Si). Large area heteroepitaxial growth of β-SiC on Si was finally achieved by NISHINO et al. at NASA Lewis using a CVD process [16]. Many groups are now using this approach to grow β-SiC single crystal films [18-20]. Over the last five years, significant progress has been made in characterizing these films, improving their quality, and in fabricating prototype devices. The following is a brief description of this process as implemented at NASA Lewis [12,21].

Growth takes place in a fairly conventional horizontal CVD system. The Si substrates are supported and heated by an rf-heated graphite susceptor. The process steps are shown in Fig. 2. Hydrogen is the carrier gas and flows continuously during the process. Other gases are added for the various steps. In the first step, HCl is added for two minutes with the Si substrate at 1200°C to

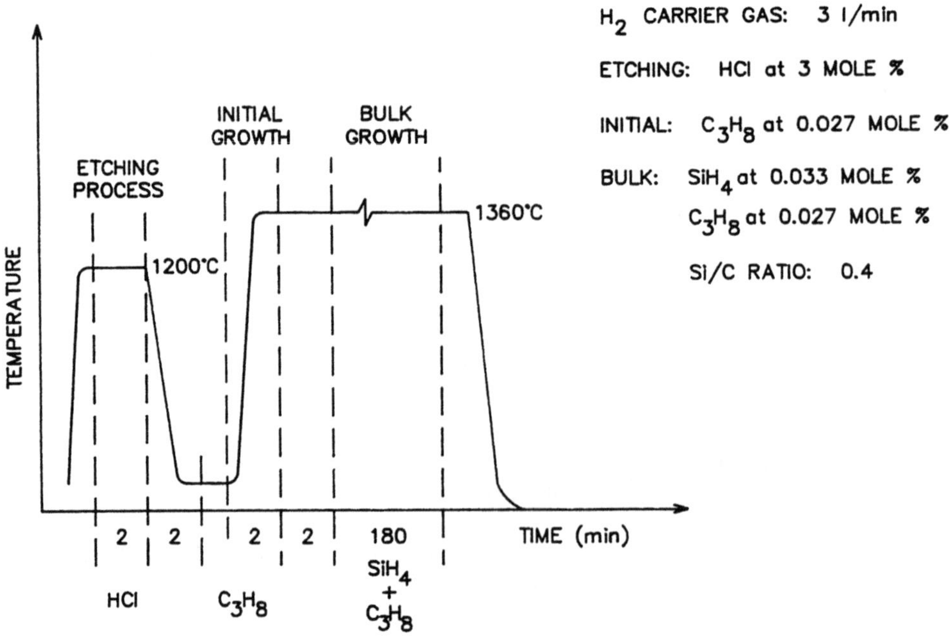

Fig. 2 Process steps for growing β-SiC on Si

chemically remove about 2 μm of the Si surface. This provides a good surface for the next step, the growth of the initial SiC layer. The essential element of this step is a rapid temperature ramp from near room temperature to the growth temperature in the presence of a hydrocarbon. To prepare for this step, the susceptor and substrates are allowed to cool for several minutes after the HCl etch. When near room temperature is reached, C_3H_8 is added to the carrier gas. After allowing an additional 30 seconds for flow equilibrium, the temperature of the susceptor and substrate is ramped to 1360°C in about 30 seconds. The C_3H_8 is left on for an additional 90 seconds and then turned off. During the 2 minute period (ramp plus 90 seconds of additional growth), a single crystal β-SiC film, about 20 nm thick, is produced on the Si substrate. After a two-minute purge with the H_2 carrier gas, SiH_4 and C_3H_8 are added to initiate the final step, the bulk growth of β-SiC to the desired thickness. With the conditions listed in Fig. 2, film growth rate is about 3.3 μm/h.

A typical 3-hour run yields a 10-μm-thick β-SiC single-crystal film on the Si substrate. The orientation of the film is the same as the substrate. A (001) substrate orientation is preferred because the (111) orientation produces films that are severely warped and cracked.

5.2 Problems with β-SiC Films Grown on Si

Extensive investigation of the β-SiC films grown on Si have indicated problems in film quality. In this section, the problems and some solutions will be reviewed.

A major problem with the β-SiC grown on Si is a high density of defects. Transmission electron microscopy (TEM) studies have identified misfit dislocations, twins, stacking faults, and antiphase boundaries (APB's) [22-24]. The first two are generally found at the interface between the SiC and Si or within a few μm of the interface [25]. However, the latter two frequently extend all the way to the surface of the film. The situation is similar to the growth of other polar compounds (e.g. GaAs [26] and GaP [27]) on Si. These defects are generally attributed to lattice mismatch and the difference in thermal expansion coefficents [23,25]. The mismatch relative to Si is -20%, +4%, +0.4% for SiC, GaAs, and GaP, repectively. The unsatisfactory performance of β-SiC prototype devices may be due to these defects.

For the problem of APB's, a solution found for the growth of GaP on Si [28] was successfully applied to β-SiC on Si [29,30]. This consists of using vicinal (001)Si substrates that are tilted about a <110> direction. For SiC, tilt angles of 1°, 2°, and 4° were successful in eliminating APB's. Also, the surface morphology was significantly improved, with a 1° tilt producing the smoothest surface. Preliminary data indicated that the use of tilted substrates did not reduce the density of stacking faults [30]. This development is discussed further by MATUS et al. in these Proceedings.

A study of the heteroepitaxy of GaP on Si suggests that the initial nucleation process, rather than lattice mismatch or differences in the expansion coefficient, may be responsible for the formation of stacking faults [31]. One observation is that the stacking fault density in such GaP films is very high despite the small 0.4% lattice mismatch. Also, high resolution cross-section TEM indicates that stacking faults may form on (111) facets that are observed on isolated three-dimensional GaP nuclei before the nuclei coalesce. Extending these ideas to SiC raises hope that a modification of the initial nucleation process can reduce the SiC stacking fault density.

Another problem with the β-SiC films is the apparent high concentration of residual donors and compensating acceptors. Hall measurements indicate that while the net electron carrier concentration in unintentionally-doped material is in the low 10^{16} cm^{-3} range, the actual donor and acceptor concentration are about 10^{18} cm^{-3} (i.e. greater than 90% compensation of the donors by the acceptors) [32,33]. Work described by BISHOP et al. in these Proceedings indicates that there is a dominant unknown donor with a binding energy of 15-20 meV in addition to isolated nitrogen with a binding energy of 54 meV. The source of the compensating acceptors is also unknown.

5.3 Heteroepitaxial Growth of β-SiC on TiC_x

In an effort to avoid problems caused by a large lattice mismatch, Parsons at the Hughes Research Laboratories has been investigating the growth of β-SiC on TiC_x substrates. With x = 0.85 the mismatch is about 0.6%. This approach has yielded single crystal β-SiC films with smooth surfaces and good electron channeling patterns [34]. One difficulty, however, is the float-zone growth of the TiC_x single-crystal substrates, which requires a temperature of 3140°C at a pressure greater than 10 atmospheres.

5.4 (Hetero)epitaxial Growth of β and 6H SiC on 6H SiC

In another effort to avoid problems caused by lattice mismatch, Davis and coworkers at NCSU have been investigating the use of α-SiC substrates obtained as a byproduct of the Acheson process. The results have been very encouraging.

The results, which are discussed by KONG et al. in these Proceedings, are summarized here. Single crystal SiC growth was achieved on the basal plane of 6H SiC [(0001)Si face and $(000\overline{1})$C face] and on surfaces inclined to the basal plane. Growth on the basal plane yielded β-SiC films with much lower defect density (compared with growth on Si). The only significant defects were double positioning (DP) boundaries. Efforts to eliminate these DP boundaries led to growth on inclined surfaces. Growth surfaces tilted 3° toward [1120] yielded 6H SiC films (rather than β-SiC) with very low defect densities.

In the past, the results by the NCSU group may have been primarily of academic interest because the 6H SiC substrates were very irregular in shape, nonhomogeneous, and were obtained by hand picking likely samples from tons of industrial grade material. However, as will be described in the next section, there is reason to believe that large-area 6H SiC single crystal substrates will be available in the near future.

5.5 Modified Sublimation Process

Very recently, the SiC research team at NCSU announced the successful implementation of a seeded-growth sublimation process [17]. Their approach is similar to a modified sublimation process developed by TAIROV et al. in the U.S.S.R. [35,36] and ZIEGLER et al. in F.R.Germany [37]. It is an improvement over the original Lely sublimation process in that only one large crystal is grown and this crystal consists of a single polytype.

The basic elements of the modified sublimation process are shown in Fig. 3, which is a schematic diagram of the configuration used by ZIEGLER et al. [37]. Nucleation took place on a SiC seed crystal located at one end of a cylindrical growth cavity. A temperature gradient was established within the cavity such that the polycrystalline SiC was at 2400°C and the seed crystal was at 2200°C. At these temperatures SiC sublimed from the polycrystalline material and condensed on the cooler seed crystal. Growth took place on the seed crystal in an atmosphere of argon at 200 Pa. Large boules of single polytype 6H SiC were grown with growth rates of about 4 mm/h and boule sizes up to 20 mm in diameter and 24 mm long.

At NCSU, 6H SiC single crystals, 15 mm in diameter and 8 mm thick, have been grown by this seeded-growth sublimation process [17]. Unintentionally-doped crystals were transparent and colorless; nitrogen-doped crystals were green. Preliminary TEM studies indicated a much lower density of line and planar defects (compared to SiC grown on Si).

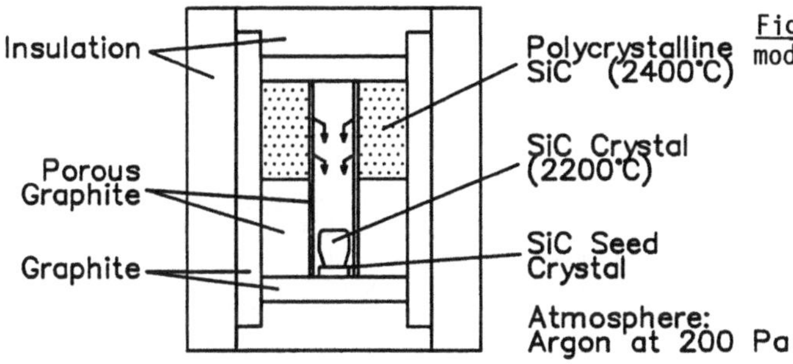

Fig. 3 Growth chamber for the modified sublimation process

With this progress, there is real hope that single crystal 6H SiC substrates will soon be available.

6. Recent Progress in Device Technology

To fabricate semiconductor devices, a variety of processes are used to introduce controlled amounts of dopants (i.e. donors and acceptors), and to oxidize, etch, and metallize the material. This section will be a review of developments in these processes.

6.1 Controlled Doping

Except for nitrogen, the choice of dopants for SiC is similar to Si; thus we have B and Al for p-type dopants and N and P for n-type. These dopants can be introduced during CVD growth [38]. Aluminum has a higher solubility than B in SiC and as an acceptor its ionization energy is less (257 meV vs 735 meV in β-SiC). So, it is becoming the preferred p-type dopant. Nitrogen is electrically active in SiC and is preferred as the n-type dopant because of its high solubility and relatively low ionization energy (54 meV in β-SiC). For further discussion of donors and acceptors in β-SiC, see BISHOP et al. in these Proceedings.

In commercial Si technology, diffusion and ion implantation are the key processes used to introduce dopants during device fabrication. Diffusion is not practical with SiC because diffusion coefficients are negligible below 1800°C. In fact, this is one of the strengths of SiC; dopants don't move at elevated temperatures.

Ion implantation has been successfully applied to SiC. Recent work by EDMOND et al. [39] demonstrated that implanting at high temperatures (approx. 550°C), followed by annealing at 1200°C, resulted in improved structural and electrical properties.

6.2 Oxidation, Etching, and Metallization

The use of oxides in SiC device fabrication is similar to Si. Deposited oxides can be used for insulation, passivation, and as masks for ion implantation. For the gate oxide in FETs, thermal oxidation produces an oxide in Si with the lowest interface trap density. Thermal oxidation has been used successfully with SiC [40], but a recent study has shown that it can cause a redistribution of dopants. For example, near the oxide/SiC interface, the p-type dopants, B and Al, are depleted, but the concentration of n-type dopants, N and P, are increased in the SiC [41].

During device fabrication, etching processes for removing portions of the semiconductor are required. In the past, molten salts and gases (e.g. Cl_2/O_2 mixture) at high temperatures served as etchants for SiC [11]. Recently, plasma-assisted etching techniques have been applied to SiC. One of the more successful applications has been reactive ion etching (RIE) on a carbon cathode using NF_3; clean etched SiC surfaces with a high degree of anisotropic etching were demonstrated [42]. Further discussion of plasma-assisted etching of SiC can be found in PAN et al. in these Proceedings.

Since the initial use of SiC devices will probably be for high temperature environments, suitable metallization for contacts is crucial. Below 400°C, the problem may not be very difficult. An alloy of Au with a few percent of Ta has been used at these temperatures [11]. At higher temperatures, a multi-layer structure will probably be required. The goal of the various layers will be to provide an ohmic contact, while also providing a diffusion barrier to prevent intermixing of materials. Various metal silicides (e.g. $TaSi_2$ [43]) and carbides (e.g. TiC [44]) are being investigated. Stoichiometry is very important in the stability of contact materials. For example, TiC plus excess C was found to be stable in contact with SiC after 500 hours at 800°C in a vacuum environment. With a deficiency of C, the TiC quickly reacted with the SiC [44].

6.3 Prototype Devices

The emphasis on β-SiC in recent years has led to the development of devices fabricated from this polytype. The most notable device to date has been a depletion-mode metal-oxide-semiconductor FET (MOSFET) produced at NCSU [43]. It has excellent high temperature characteristics up to 650°C. The substrate for this β-SiC device was an industrial-grade 6H SiC crystal. Devices produced at NCSU from β-SiC grown on Si substrates include inversion-mode MOSFETs, MESFETs, ion-implanted diodes, and Schottky diodes. These devices are described by KONG et al. in these Proceedings.

KELNER et al. [45] have reported on Schottky diodes and MESFETs developed at NRL. AVILA et al. [46] have characterized ion-implanted diodes fabricated from β-SiC grown on Si at NASA Lewis.

Second harmonic generation was recently demonstrated at Howard University in β-SiC grown on Si. This effect, which may lead to the development of SiC optical devices, is described by JACKSON et al. in these Proceedings.

7. Decisions and Challenges Ahead

We believe that SiC developments in the USA during the last year have marked a turning point for this unique semiconductor. The work at NCSU demonstrated the feasibility of producing large 6H SiC substrates and also the feasibility of making excellent high temperature SiC devices. It is no longer a question "if" SiC will be successful, but "when". Because of limited research funds, decisions will have to be made as to which growth process, which polytype, and which devices to pursue.

At this time, the performance of prototype devices that use 6H SiC as a starting material are much superior to those using β-SiC grown on Si. Hence, the emphasis in CVD growth for device fabrication will probably shift to using 6H SiC substrates. Much work remains in the development of the seeded-growth sublimation method that will supply these substrates, and in characterizing the resulting material. Research on methods of growing β-SiC on Si should continue because of the potential economic advantage of starting with inexpensive Si substrates.

Beta SiC and 6H SiC each have advantages over the other. For example, the simpler structure of β results in it having a larger electron mobility and an advantage for high frequency devices. The larger bandgap of 6H SiC allows it to be used for blue LEDs, and for operation at higher temperatures than β-SiC. Hence, there are reasons for pursuing both polytypes.

As interest in SiC devices picks up, there must remain a balance between basic materials research and device research. If research shifts too much toward strictly devices, then history will have repeated itself. In the 1960's, too much emphasis was placed on developing specific devices before the material was ready. Some difficult challenges lie ahead in understanding the growth of SiC (e.g. the source of structural defects and electrical compensation).

As in the development of Si and GaAs, SiC semiconductor technology will not mature overnight. But, considering the recent SiC developments, initial benefits of SiC semiconductor devices should soon be realized.

References

1. R. Verma and P. Krishna: Polymorphism and Polytypism in Crystals (Wiley, New York 1966)
2. D. Pandey and P. Krishna: in Current Topics in Materials Science, ed. by E. Kaldis, Vol. 9, (North-Holland Publishing Co., Amsterdam 1982) Chap. 2, pp.415-491
3. A. Johnson: RCA Review 26, 163 (1965)
4. R. W. Keyes: in Silicon Carbide 1973, ed. by R. C. Marshall, J. W. Faust, and C. E. Ryan (University of South Carolina Press, Columbia, South Carolina 1974) p.534
5. W. F. Knippenberg: Philips Res. Repts. 18, 161 (1963)
6. W. J. Choyke: in Proc. Mats. Res. Soc. Symp., Vol. 97, 207 (1987)
7. J. A. Lely: Ber. Dt. Keram. Ges. 32, 229 (1955)
8. Silicon Carbide, A High Temperature Semiconductor, ed. by J. R. O'Connor and J. Smiltens, (Pergamon, New York, 1960)
9. Silicon Carbide-1968, ed. by H. K. Henisch and R. Roy, (Pergamon, New York, 1969)
10. Silicon Carbide-1973, ed. by R. C. Marshall, J. W. Faust, and C. E. Ryan, (University of South Carolina Press, Columbia, South Carolina, 1974)
11. R. B. Campbell and H. Chang: in Semiconductors and Semimetals, Vol. 7B, (Academic Press, New York 1971) Chap. 9, pp. 625-683
12. J. A. Powell: in Proc. Mats. Res. Soc. Symp., Vol. 97, 159 (1987)
13. W. E. Nelson, F. A. Halden, and A. Rosengreen: J. Appl. Phys. 37, 333 (1966)
14. L. B. Griffiths and A.I. Mlavsky: J. Electrochem. Soc. 111, 805 (1964)
15. W. F. Knippenberg and G. Verspui: Philips Res. Repts. 21, 113 (1966)
16. S. Nishino, J. A. Powell, H. A. Will: Appl. Phys. Lett. 42, 460 (1983)
17. C. H. Carter,Jr., L. Tang, and R. F. Davis: presented at the Fourth National Review Meeting on the Growth and Characterization of SiC, Raleigh, NC (1987)
18. A. Addamiano and J. A. Sprague: Appl. Phys. Lett. 44, 525 (1984)
19. K. Sasaki, E. Sakuma, S. Misawa, S. Yoshida, and S. Gonda: Appl. Phys. Lett. 45, 72 (1984)
20. P. Liaw and R. F. Davis: J. Electrochem. Soc. 132, 642 (1985)
21. J. A. Powell, L. G. Matus, and M. A. Kuczmarski: J. Electrochem. Soc. 134, 1558 (1987)
22. C. M. Chorey, P. Pirouz, J. A. Powell, and T. E. Mitchell: in Semiconductor-Based Heterojunctions: Interfacial Structure and Stability, ed. by M. L. Green, J. E. E. Baglin, G. Y. Chin, H. W. Deckman, W. Mayo, and D. Narasinham, (The Metallurgical Society, Inc., Warrendale, PA 1986) pp. 115-125
23. S. R. Nutt, D. J. Smith, H. J. Kim, and R. F. Davis: Appl. Phys. Lett. 50, 203 (1987)
24. P. Pirouz, C. M. Chorey, and J. A. Powell: Appl. Phys. Lett. 50, 221 (1987)
25. J. Ryu, H. J. Kim, and R. F. Davis: Appl. Phys. Lett. 47, 850 (1985)
26. M. Akiyama, Y. Kawarada, and K. Kaminishi: J. Cryst. Growth, 68, 21 (1984)
27. K. Morizane: J. Cryst. Growth, 38, 249 (1977)
28. M. M. Al-Jassim, J. M. Olson, and K. M. Jones: in Proc. Mats. Res. Soc. Symp., Vol. 62, 49 (1986)
29. K. Shibahara, T. Saito, S. Nishino, H. Matsunami: I.E.E.E. Electron Dev. Lett. EDL-7, 692 (1986)
30. J. A. Powell, L. G. Matus, M. A. Kuczmarski, C. M. Chorey, T. Cheng, and P. Pirouz: Appl. Phys. Lett. 51, 823 (1987)
31. F. Ernst and P. Pirouz: Submitted to J. Appl. Phys. (1988)

32. A. Suzuki, A. Uemoto, M. Shigeta, K. Furukawa, and S. Nakajima: Appl. Phys. Lett. 49, 450 (1986)
33. B. Segall, S. A. Alterovitz, E. J. Haugland, and L. G. Matus: Appl. Phys. Lett. 49, 584 (1986)
34. J. D. Parsons: in Proc. Mats. Res. Soc. Symp., Vol. 97, 271 (1987)
35. Y. M. Tairov and V. F. Tsvetkov: J. Cryst. Growth 43, 209 (1978)
36. Y. M. Tairov and V. F. Tsvetkov: J. Cryst. Growth 52, 146 (1981)
37. G. Ziegler, P. Lanig, D. Theis, and C. Weyrich: I.E.E.E. Trans. Electron Devices ED-30, 277 (1983)
38. H. J. Kim and R. F. Davis: J. Electrochem. Soc. 133, 2350 (1986)
39. J. A. Edmond, S. P. Withrow, W. Wadlin, and R. F. Davis: in Proc. Mats. Res. Soc. Symp., Vol. 77, 193 (1987)
40. C. D. Fung and J. J. Kopanski: Appl. Phys. Lett. 45, 757 (1984)
41. J. W. Palmour, R. F. Davis, H. S. Kong, S. Corcoran, and D. Griffis: presented at the Fourth National Review Meeting on the Growth and Characterization of SiC, Raleigh, NC (1987)
42. J. W. Palmour, R. F. Davis, P. Astell-Burt, and P. Blackborow: in Proc. Mats. Res. Soc. Symp., Vol. 76, 185 (1987)
43. J. W. Palmour, H. S. Kong, and R. F. Davis: Appl. Phys. Lett. 51, 2028 (1987)
44. J. J. Bellina, Jr. and M. V. Zeller: in Proc. Mats. Res. Soc. Symp., Vol. 97, 265 (1987)
45. G. Kelner, S. Binari, and K. Sleger: in Proc. Mats. Res. Soc. Symp., Vol. 97, 227 (1987)
46. R. E. Avila, J. J. Kopanski, and C. D. Fung: J. Appl. Phys. 62, 3469 (1987)

6H-SiC Studies and Developments at the Corporate Research Laboratory of Siemens AG and the Institute for Applied Physics of the University in Erlangen (FRG)

P.A. Glasow

Corporate Research Laboratory Siemens AG,
D-8520 Erlangen, Fed. Rep. of Germany

1. Historical Review of 6H-SiC Activities in Europe (Fig. 1)

The beginning of SiC optoelectronic devices can be traced to Europe: In 1907 Round from the Marconi company observed SiC electroluminescence /1/ for the first time.

Then for a long time very little happened. In 1955, Lely /2/ published his crystal growth method. In 1959, 1968 and 1973 three international conferences on silicon carbide took place, the last in Miami Beach. This movement reflects the general trend of searching for semiconductors with a wider bandgap than Si and Ge at that time. One result of that trend was the increasing activity on III-V-compounds, initiated by Professor Welker at our Siemens Research Laboratory in Erlangen, Germany in 1953. In 1963 Knippenberg from Philips in Eindhoven reviewed crystal growth phenomena in silicon carbide.

Leningrad	Joffe Institute Uljanov Institute
Eindhoven	Philips
Amsterdam	University *van Daal et al*
London	*Brander*
Sofia	*Lilov*
Dresden	Akademie der Wissenschaften
Hannover	*von Münch et al*
Erlangen	Siemens University
Cardiff University	Department of Appl. Physics and Electronics,
München	Messerschmidt–Bölkow–Blohm

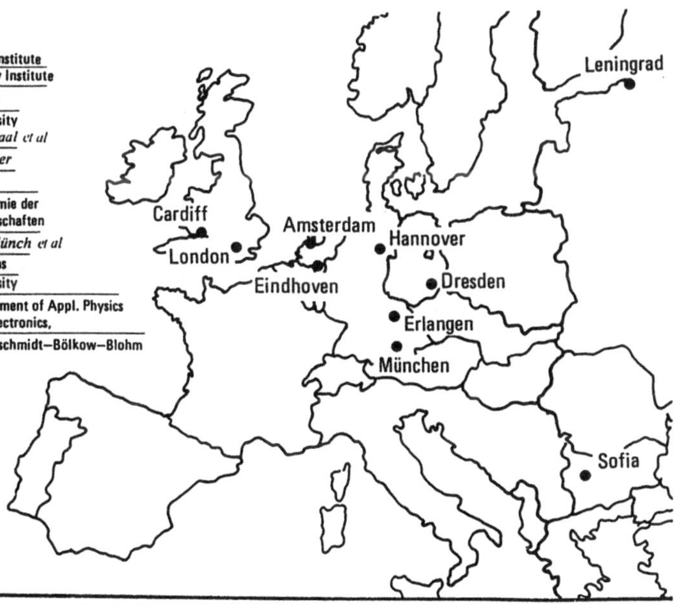

Fig. 1: 6H–SiC activities in Europe

After 1973 worldwide work on SiC slowed down appreciably. However, reports still came from Westinghouse (Pittsburgh, USA) /3/, General Electric (Cleveland), and from Kyoto University (Japan) /4/.

European companies and institutes also geared down their work on SiC during that time until 1980. However, van Daal (University of Amsterdam) /5/ and Münch (FRG) /6/ published results on SiC, and Köninger /7/ at MBB used SiC diodes as high energy radiation detectors. A number of publications about SiC activities came from the USSR /e.g. 8/, including papers about solid state properties and sublimation processes. In 1977, Tairov and Vodakov /9/ published an excellent survey about SiC activities. All European groups have worked on 6H-SiC only, up to now, with the exception of some USSR groups, which have also published about 4H and 3C material.

After 1980 interest in 6H-SiC rapidly increased in Europe, especially in Leningrad, USSR, and in Erlangen, FRG. The reasons for these new activities were:

1. progress in single crystal growth of 6H modification,
2. progress in device technology, especially in ion implantation, etching, and photo pattern techniques.

Today, in the USSR up to 21 labs are working on SiC; among these are the Ulyanov Institute (Tairov) and the Joffe Institute (Vodakov), both in Leningrad. One interesting, recent development of the Joffe Institute is the three color blue-green-red display made from one single crystal /10/.

In Erlangen, F.R. Germany, present activities include

1. Activities at the Institute of Applied Physics, University of Erlangen:
More basic research related work about physical material characterization using Hall measurements, DLTS, PL and EL, but also technology, primary ion implantation on material grown at the Siemens Research Laboratories.

2. Activities at the Siemens Corporate Research Laboratory itself:
More device related work where single processing steps, like the Lely process, material characterization with regard to the modification, crystal imperfections like EPD, LPE and chip technology were improved and as first optoelectronic devices the blue LED and an UV sensitive photodetector have been developed.

2. SiC-Crystal Growth Methods

In principle, Al doped SiC ingots of the 6H modification are grown by 3 different methods:

1. Acheson uncontrolled sublimation method for bulk crystals, which results in highly (up to 10^{19} cm^{-3}) contaminated and highly compensated material with low mobility, e.g. about 30 $cm^2V^{-1}sec^{-1}$ for electrons.

2. The better controlled sublimation process for bulk crystals is the modified Lely.
The material is still highly compensated, however, the impurity concentration in undoped material is lower and electron mobilities up to 800 cm^2/Vs can be obtained.

3. The liquid phase epitaxial method, which is used for the growth of
a) active layers purer than the substrate material: doping one order of magnitude lower than in substrates can be obtained, and b) to form pn-junctions by overcompensation of Al acceptors with N-donors.

2.1 The Acheson Process

In the large-scale technical production of silicon carbide for technical purposes as grinding powders, Acheson's method /11/ is still used in its essentially unaltered form.

Fig. 2 shows a piece of such Acheson grown SiC. The SiC platelets are clearly to be seen. They have mostly the 6H, 4H or 15R modification.

Fig. 2: SiC druse with platelike crystals produced by the Acheson process
(Elektroschmelzwerk Kempten, F.R.Germany)

2.2 Modified Lely Process Using Seed Crystals

The original Lely method /2/ produced platelet monocrystals. From 1955 to 1980, these platelets were used for fundamental investigations.

However, to grow 6H-SiC ingots large enough to be technically useful as substrate material the best way is the sublimation growth method with controlled deposition of SiC on a seed crystal in a temperature gradient, the modified Lely method /12,13/. The three essential parameters for crystal growth are growth temperature, temperature gradient, and pressure inside the reaction chamber. The growth temperature together with the polytype and the orientation of the crystal used mainly determines the polytype of the growing crystal, whereas temperature gradient and gas pressure control the transport and hence the crystal growth velocity.

The schematic view of the growth apparatus is shown in Fig. 3.

The reaction chamber consists of the outer cylinder of electrographite and the inner cylinder of porous graphite. The lower part of the space between these two cylinders is filled with porous graphite, whereas the upper part contains Al, N doped SiC powder, supplied by Elektroschmelzwerk Kempten. The porosity of the inner graphite cylinder allows material transport from the supply chamber to the inner growth chamber. The crucible assembly is covered by a heat shield (Sigraflex).

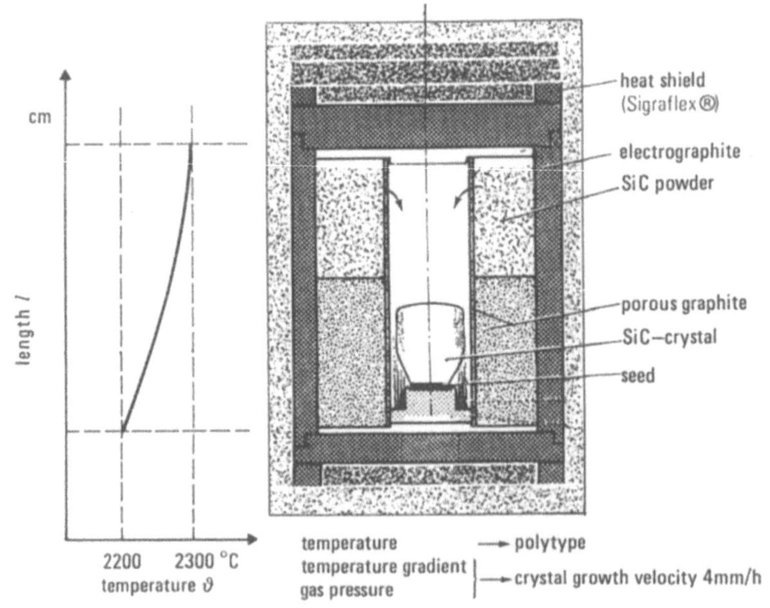

Fig. 3: Crucible assembly for the growth of SiC single crystals by the modified Lely method

Heating is performed by RF induction (4 kHz, 20 kW). The temperature profile inside the chamber is measured with a pyrometer through viewpoints in the top and bottom covering plates.

A negative temperature gradient of 10° to 25°C/cm^{-1} towards the seed crystal is adjusted in contrast to the conventional Lely method. The seed crystal (Acheson material or crystals grown in our laboratory) usually of 14-mm diameter and 0.4-mm thickness is mounted on top of a platform about 20 mm below the coldest part of the SiC source material.

Typical temperature gradients which yielded good results were around 15°C/cm, resulting in a crystal growth velocity of about 4 mm/h.

Fig. 4 shows several Lely grown crystals.

Fig. 5 shows a cross-sectional view of one of our early crystals grown on an Acheson seed. The dark part at the bottom is the seed crystal. A large number of small pores with tiny bubbles at their ends extend from the top of the seed into the crystal. Their occurrence is interpreted as follows: The surface of the Acheson seed crystal exhibits a large number of defects consisting of small holes. When crystal growth starts small cavities are formed in which, due to the temperature gradient, material transport occurs from the upper region to the lower region of the small cavity. Consequently these cavities move upwards leaving behind them the traces which can be seen in Fig. 7.

Fig. 4: 6H SiC crystals grown by the modified Lely method

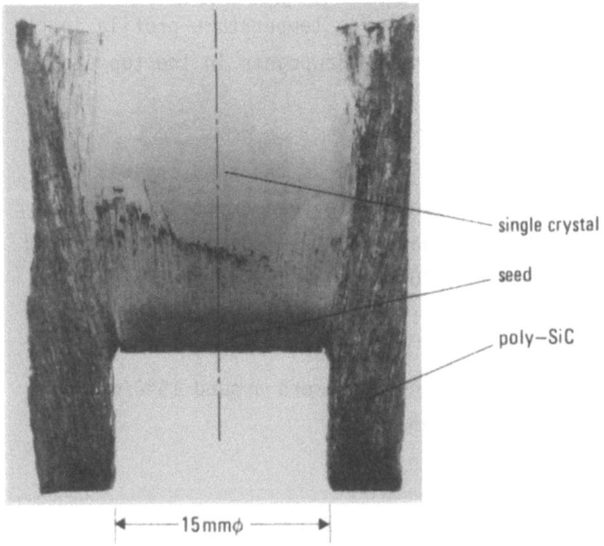

Fig. 5: Cross–section of a 6H–SiC ingot, exhibiting polycrystalline coating and pores reaching into the crystal

We grow with our standard procedure (2200°C, 15°C/cm, 2 mbar Ar) 6H-SiC single crystals with 14-20 mm useful diameter and up to 24 mm useful length within 8 h including the loading procedure. In over 200 runs we obtained a yield of 80 percent for the desired 6H-modification.

2.3 SiC Liquid-Phase Epitaxy

Since the LPE-process is well known from other processes, e.g. manufacturing LED's (blue light emitting diodes), and since the SiC-LPE process is well covered in the literature /e.g. 14/, it does not require lengthy discussions.

In principle the epitaxial growth of the SiC layer is carried out in a graphite crucible, filled with a carbon saturated silicon melt. Carbon dissolved from the crucible wall automatically causes silicon saturation of the melt. The layer growth takes place lowering the temperature from a starting point around 1600°C. Usually, layers with about 5 μm thickness are grown.

3. Material Characterization

3.1 Chemical Purity

The chemical purity of the base material and of the Lely grown crystals can be investigated by spectral analysis. This permits impurities to be detected, but is

contamination	base–material	run			
		2	4	5	8
Ag	–	–	–	0.5	–
Al	180	70	70	40	50
Ca	–	–	–	–	–
Cu	0.15	–	–	–	–
Fe	140	–	–	–	–
Mg	66	2.6	1.2	1.3	1.6
Ni	19	–	–	–	–
Ti	50	25	15	10	24
V	50	40	16	11	30

Fig. 6: Contamination of base material and Lely grown crystals measured by spectral analysis [ppm]

also indicative of their origin and their transportation during the process. Fig. 6 shows a list of analysis results of base material and of Lely grown crystals.

The V and Ti contaminations don't originate from an abrasive effect of any crashing tools e.g., but are impurities in the quartz sand. Their concentrations depend on the location where the sand is mined. The same applies to carbon. The main impurities in the Acheson starting material are Al, Ca, Mg, Fe, Ti, V, Ni.

3.2 Etch Pit Density

The etch pit density in sublimation grown crystals is about 10^4 cm^{-2} /15/.

3.3 Morphology Determination by Absorption Measurements

A permanent problem in SiC-processing and technology is the polytypism of the material. We have found that single SiC wafers can contain more than one of the various polytypes (2H, 4H, 6H, 15R, 3C). For the device to be built on the wafer, however, usually a particular polytype is required. Related to LED's, e.g. the color of the emitted light is determined by the modification. Therefore, for reasons of yield and costs it is highly desirable to identify regions of unwanted polytypes prior to any further processing steps e.g. epitaxy. This identification can be done optically in 2 ways based on the different bandgaps of the different polytypes.

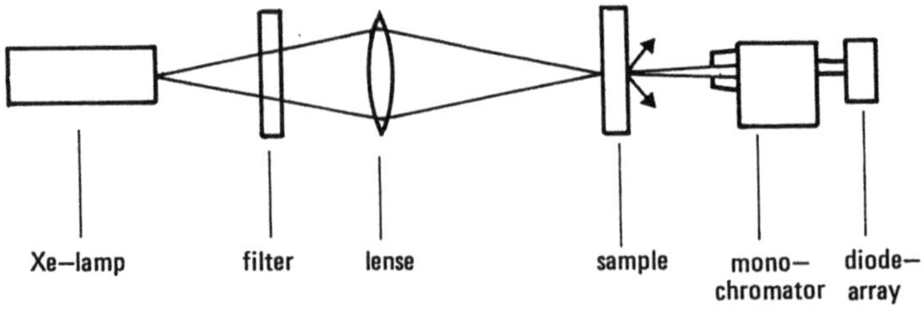

Fig. 7: Schematics of experimental setup for determination of SiC–modification by transmission measurement

The first method (Fig. 7) uses illumination from a broad band Xenon light source (350 to 650 nm). The detection of the spectral composition of the transmitted light allows one to visualize the absorption of the material which is sufficient to characterize the structure. An experimental setup is shown schematically. A xenon lamp in combination with a set of optical filters produces a light of a broad spectral composition. This light is focussed onto the sample (spot size 0.3 x 0.3 mm^2) and the transmitted light is detected by an optical multichannel analyser in combination with a 20 cm monochromator. The xenon lamp spectrum and typical transmission spectra for two 4H and 6H SiC structures grown by the Lely method are plotted in Fig. 8. Additionally, a transmission spectrum of 6H Acheson material is shown. One clearly distinquishes the 4H and 6H structure showing an absorption edge at 380 (400) and 420 (427) nm, respectively. The spectrum at larger wavelength is similar to the lamp spectrum indicating no or only small concentrations of dopants or impurities within the bandgap. Contrary to the Lely substates the Acheson material exhibits in addition to the 6H absorption edge strong absorption

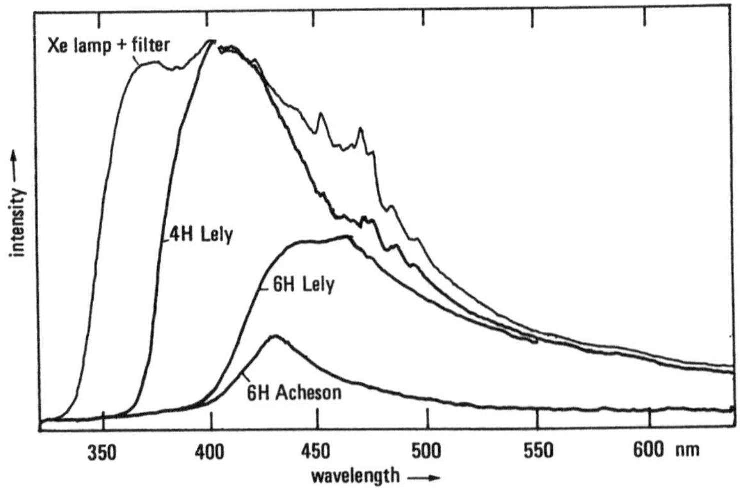

Fig. 8: Spectrum of lamp/optical filter combination and typical transmission spectra of 4H, 6H Lely and 6H Acheson grown crystal [25]

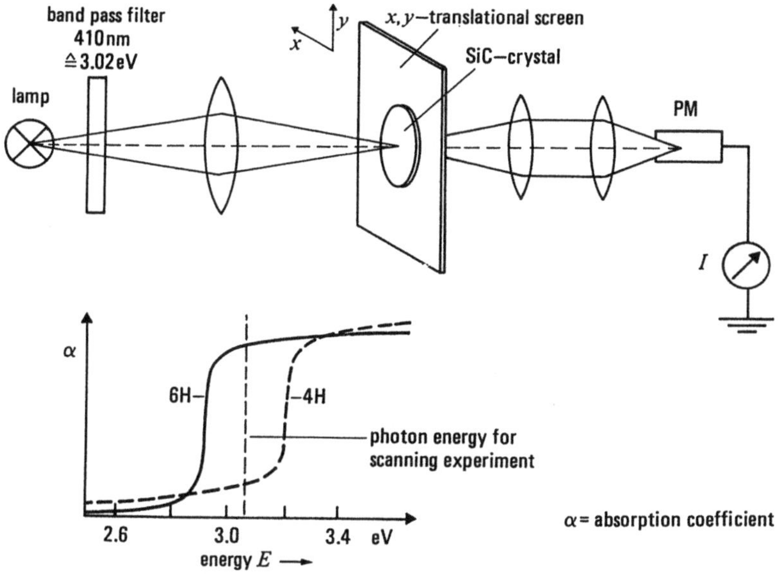

Fig. 9: Experimental setup for transmission topography with a certain photon energy between the 6H– and 4H–SiC bandgap

over the whole lamp spectrum, indicating a broad distribution of impurities in the forbidden gap.

Principle (lower part) and experimental scheme (upper part) of the second method for a discrimination between regions of 6H and 4H on a same Lely substrate are shown in Fig. 9. The absorption edge of 6H is at 2.9 eV (lower part of figure), the one of 4H at 3.2 eV. Here transmission topography (upper part) with a certain photon energy between the two bandgaps, selected by the bandpass filter as 3.02 eV corresponding to 410 nm, is used. Fig. 10 shows the result of a transmission scan of a Lely SiC-wafer with 4H and 6H regions. The wafer appears dark in its 6H regions, and transparent in its 4H regions. The dark regions correspond to 6H, the light ones to 4H modification.

3.4 Electrically Active Impurities Detected by Hall Measurements (Fig. 11).

In spite of a large scatter of the results of the Hall measurements, the following facts were found as shown in typical results:

a) At room temperature the resistivity of Lely grown material is about one order of magnitude lower than that of Acheson material, resulting from the higher purity and larger mobility of the charge carrier.

b) Acheson material is p-type doped with an Al-concentration close to 10^{19} cm^{-3}. The impurity concentration in Lely material is one order of magnitude lower than the corresponding concentration in the purest Acheson samples. Lely material is n-type with an impurity concentration around 1 · 10^{18} cm^{-3}. At room temperature Lely material shows a charge carrier concentration 3 · 10^{17} cm^{-3}. m_{eff}/m_0 has been taken from /28/.

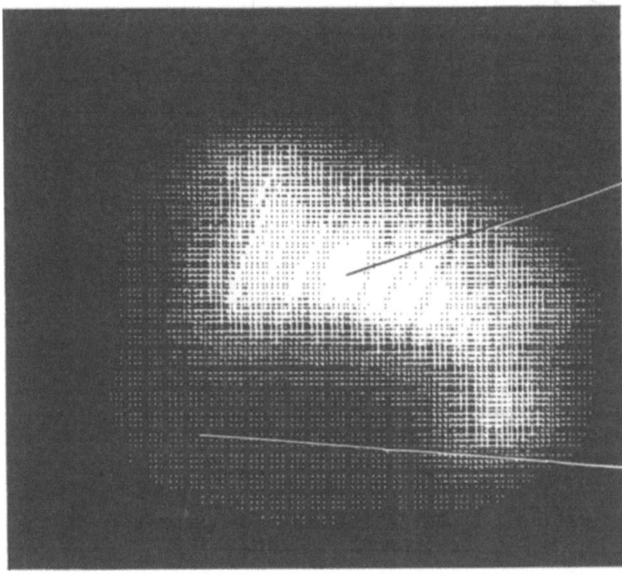

Fig. 10: Absorption–Mapping of a SiC sample to separate 4H and 6H modification [26]

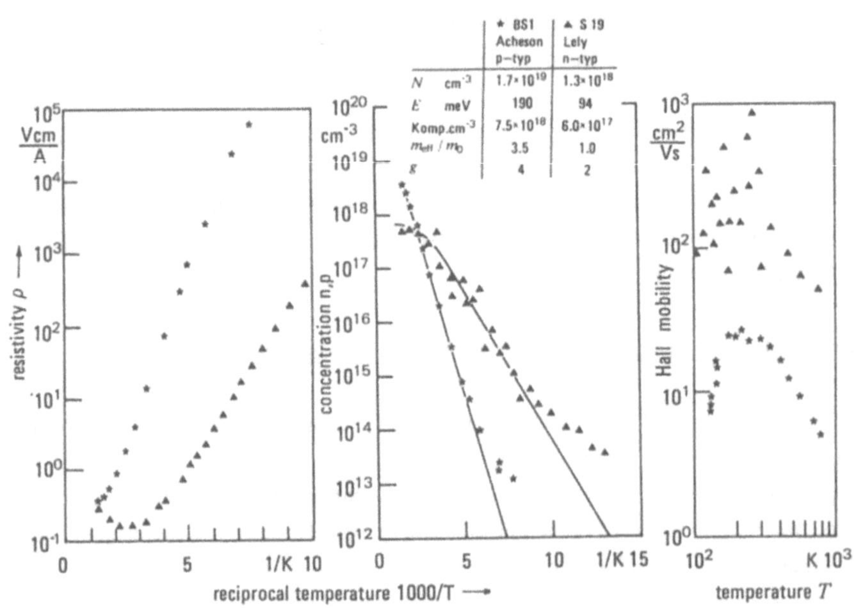

Fig. 11: Resistivity ϱ, charge carrier concentration (n,p), and Hall mobility of samples BS1 (SiC:Al (Acheson)) and S19 (SiC:N, Lely) as function of temperature [16,17]

c) Acheson p-type material shows a maximum hole mobility of 25 cm^2/Vsec at T = 200 K, whereas the electron mobility in n-type Lely material has been measured as around 800 cm^2/Vsec.
The scatter of the results in Fig. 11 is caused by the quality of the crystals and by the influence of nonohmic contacts (Ni).

The rise in mobility with decreasing temperature can be ascribed to a decrease of phonon scattering; the drop in mobility as temperature decreases further can be traced to scattering at ionized impurities and impurity band conduction.

These measured values of the electron mobility in Lely grown SiC correspond to the best values given in the literature. A direct comparison of electron mobilities in Lely and in Acheson material is not possible, since there is no "as grown" n-type Acheson 6H-SiC. However, we can compare n-type implanted layers in p-type Acheson or Lely material. The comparison with electron mobilities in n-implanted layers shows that the electron mobility in Lely substrate is greater by a factor of 3 /17/.

3.5 Photoluminescence

In the PL-spectra of Lely material the familiar Ti lines can clearly be seen (Fig. 12). The lines are very narrow, indicating crystals of high quality /27/.

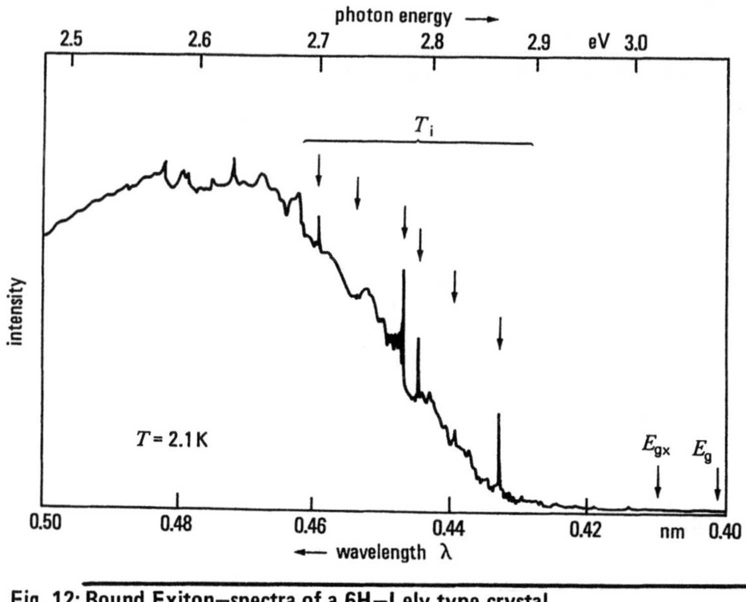

Fig. 12: Bound Exiton–spectra of a 6H–Lely type crystal. Excitation by Krypton–laser 150 mW (351 nm, 356 nm)

3.6 Deep Level Transient Spectroscopy

DLTS-measurements were done on Ti-Schottky contacts with a computer controlled DLTS arrangement /18,19/.
In p-type Lely crystals an unknown level at 457 meV above valence band has been found (Fig. 13), whereas in Acheson material an also unknown impurity has been found at 377 meV above valence band (Fig. 14). No other levels could be observed by DLTS.

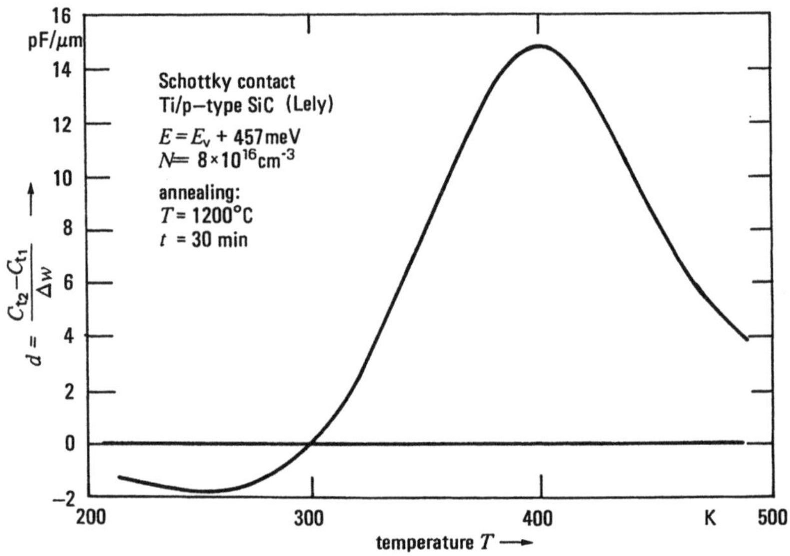

Fig. 13: Typical DLTS spectrum of a Lely grown 6H–SiC sample [19]

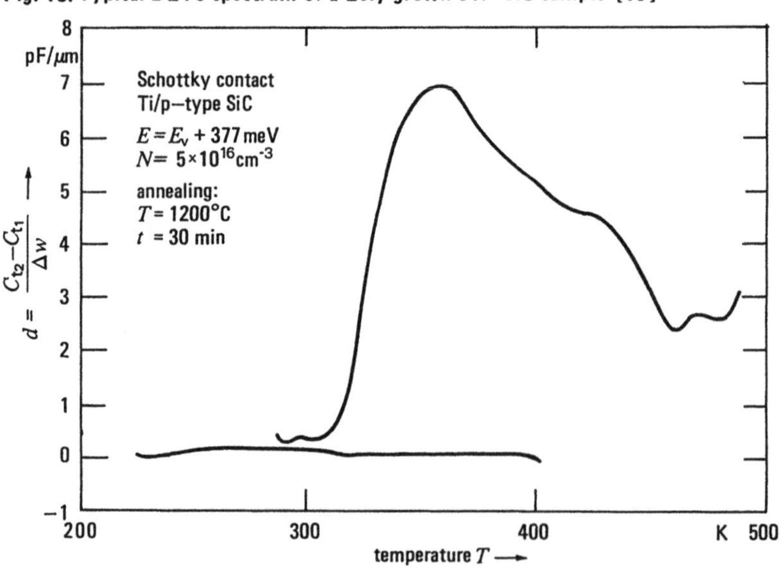

Fig. 14: Typical DLTS spectrum of an Acheson grown 6H–SiC sample [19]

4. Device technology

After base material improvement, the question arose whether unipolar or bipolar devices should be developed.
Because of the still relatively high impurity concentration in the substrate material, it was obvious to start with bipolar devices. Unipolar devices require lower impurity concentration.

However, the difficulty in developing bipolar devices was the formation of the pn-junction. In a first step, we have developed pn-junctions with good performance in the forward direction on the basis of N-doping in p-type epitaxial layers. This was the first stage in the development of blue light emitting diodes. However, this pn-junction has a relatively poor performance in the reverse direction.
We have achieved in a second step pn-junctions with high reverse voltage and low current by /14/ N-implantation and RIA processes in comparison with furnace annealing. This development also served as the approach to the planar technology of SiC. The low annealing temperatures used in RIA do not effect the SiO_2-layers, thereby allowing the usual structuring in planar technique.

4.1 Blue Light Emitting SiC: Al, N Diodes

The fabrication technology for blue light emitting diodes is essentially a liquid phase epitaxial method where the p-n junction is formed by overcompensation of Al acceptors with N donors, and is described in /14/.

4.1.1 Degradation phenomena in blue-light-emitting SiC diodes

One of the most important parameters of semiconductor devices is their long term reliability, and hardly any degradation at all has been observed over more than 15000 hours with blue-light-emitting SiC diodes /20/. However, unexpectedly we found that some of the blue-light-emitting SiC diodes changed their emission color from blue to greenish-blue within the first minutes of operation, even under low current density conditions /21/. This color change was accompanied by a decrease in quantum efficiency.

We found out that in our degraded diodes of a 6H-polytype SiC stacking faults are formed which migrate from centers of enhanced nucleation at the sawed edges into the crystal. This degradation could largely be avoided by etching and careful handling of the diode chips.

4.2 Implanted 6H-SiC UV-Photodiodes

In a second step we have developed photodiodes on the basis of 6H-SiC optimized for detection of UV radiation /22/.

In comparison to other photodetectors, sensitive in the near UV region, SiC photodiodes offer some important advantages, as summarized in Table 1.

In contrast to Si-photodiodes and to photomultipliers which are sensitive in a large spectral range (e.g. in the case of Si from 200 to 1150 nm), the responsivity of SiC detectors is restricted to the spectral range from 250 to 450 nm. Therefore, UV detectors based on SiC do not require expensive blocking filters to limit the detector sensitivity to the UV spectral range. In addition, the UV-responsivity of SiC detectors is superior to those of photocathodes and comparable to those of Si devices. A further advantage of SiC detectors is the fact that the absorption coefficients below 400 nm are more than two orders of magnitude smaller than the absorption coefficients of Si. Consequently, surface recombination effects should be less important in SiC than in Si photodiodes related to comparable junction depths. Finally SiC photodiodes can also in principle operate up to temperatures of 1100°C without degradation and also their radiation resistance should be superior to any other semiconductor material.

In the following, the fabrication of implanted n^+p-junctions is described as well as the characterization of those SiC diodes by their I-V characteristics and by their quantum efficiencies at temperatures up to approx. 400°C.

Table 1: Properties of different UV – photodetectors

	UV – Photodetectors		
	SiC–Photodiode	Si–UV–Photodiode	Photocathode (Multiplier etc.)
wavelength range	$200nm \leqslant \lambda \leqslant 450nm$	$200nm \leqslant \lambda \leqslant 1150nm$	$200nm \leqslant \lambda \leqslant 600nm$
antiblooming for visible light	without blocking filter	blocking filters required	blocking filters required
responsivity in A/W for λ=280 nm 300 K	0.14 (at 700 K)	0.18	0.04
design	monolithic	monolithic	complex
high temperature application	up to 700 K	no	no
radiation hardness	high (expected)	low	low

Photodiode design: SiC parameters, light absorption and n^+p-junction depth

SiC photodiodes are sensitive in the ultraviolet range from 200 to 400 nm. For these wavelengths, the optical penetration depth ranges from less then 10^{-5} to 10^{-1} cm /23/. At 300 nm and room temperature, the absorption coefficient in SiC is $5 \cdot 10^3$ cm^{-1}, corresponding to a penetration depth of 2 µm. At 250 nm, an absorption coefficient about 10^5 cm^{-1} is obtained resulting in a penetration depth of 0.1 µm. An estimate for the diffusion length of electrons (holes) at room temperature based on a mobility of 100 cm^2/Vs (20 cm^2/Vs) and a minority carrier lifetime of approx. $2 \cdot 10^{-8}$ s /11/ results in 2 µm (1 µm) diffusion length. Taking these data into account, we decided to perform a shallow N-implant with ion energy of 30 keV and a dose of 10^{14} cm^{-2}.

Fabrication of UV-photodetectors

The following process steps (Fig. 15) were employed: Al doped, p-type SiC wafers with 15 mm diameter, 300 µm thickness from modified Lely- or Acheson-grown single crystals are taken as substrates for a liquid phase epitaxial process.
An epitaxial layer of approx. 5 µm thickness is grown. Then an oxide film (thickness 300 nm) is grown on the epitaxial layer. Large area windows of 70 mm^2 or small area windows of 6 mm^2 size, are plasma etched into the oxide.

n^+p-junctions were formed by nitrogen implantation (dose 10^{14} cm^{-2}, U = 25 kV) through the oxide windows. The implanted nitrogen profile corresponds to a Gaussian distribution with a peak concentration of $2 \cdot 10^{19}$ cm^{-3} located at a depth of 0.05 µm. Annealing of the implantation damage is conducted either in a standard furnace or by rapid isothermal annealing (RIA). Based on Hall effect results, annealing in a standard furnace proved to be optimal at 1470°C for 7 min. At this elevated temperature, however, the oxide mask is largely vaporized. The advantages of the RIA process are

1) that an electrical activation of 90 % of the implanted nitrogen concentration is already achieved at a temperature of 1050°C and an annealing time of 4 min, and

2) that at these annealing conditions the undesired compensation (Al concentration) is activated to a lower degree as observed with furnace annealing.

Additionally, the oxide mask serving also as passivation of the n^+p-junction is so completely preserved.

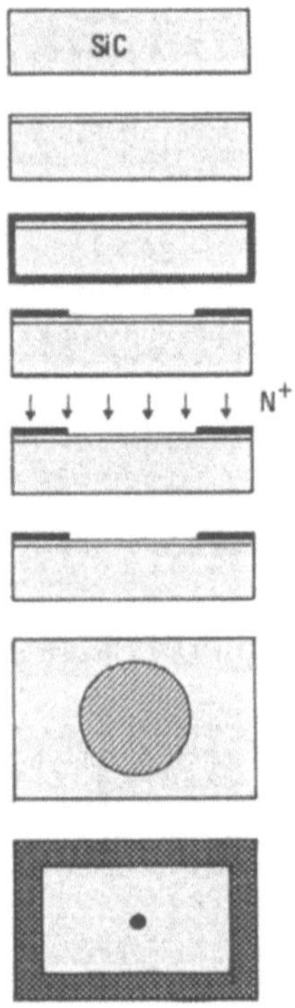

1. grinding and polishing
2. epitaxial layer growth (LPE) Al doped $|N_A - N_D| = 2 \times 10^{18}\,cm^{-3}$
3. wet oxidation
4. oxide window ($70\,mm^2$) opened by planar technique
5. nitrogen implantation ($1 \times 10^{14}\,cm^{-2}$, 25 keV)
6. RIA annealing (1050°C, 4 min) or furnace annealing (1450°C, 7 min, Si coated graphite crucible)
7. back contact Al metallisation by evaporation, RIA annealing (950°C, 15 min)
8. frontside contact (Pd metallisation)
9. packaging, bonding
10. antireflection coating (Al_2O_3, ~430 nm)

Fig. 15: SiC–UV–photodiode processing steps

Ohmic Al back contacts on the p-type substrate are prepared by evaporation and by heating the samples in a furnace (nitrogen ambient) at 950°C for 15 min. Point front contacts (diameter Ø = 100 um) are performed by Pd evaporation. Subsequently, the surface is provided with an antireflection coating (Al_2O_3).

In Fig. 16, I-V dark characteristics of a small-area SiC-photodiode at device temperatures from 296 to 826 K are plotted.

With our present knowledge, a quantitative description of the shape of the I-V characteristics is not possible. Leakage currents at reverse bias voltages and series resistances attributable to substrate or contact resistivity at higher values of forward current may control the I-V characteristics. The ideality factor decreases from 2.8 to 1.6 with increasing temperature. The ratio of forward to reverse current at room temperature and at a bias of ± 5 V is $3 \cdot 10^4$.

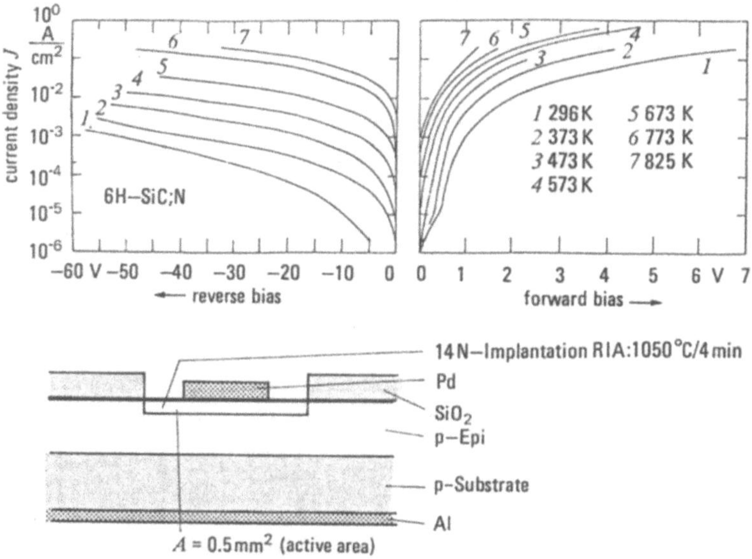

Fig. 16: *I–U* **characteristics of a N–implanted p–n diode operated at elevated temperatures [17]**

Quantum efficiency measurements

Figure 17 shows the block diagram of the experimental setup used to measure the quantum efficiency at various temperatures.

The photodiode is mounted on a Cu block which is fixed in a quartz tube. The Cu block is heated by hot nitrogen gas, flowing through the quartz tube positioned in a temperature-controlled furnace. The device temperature is monitored by a platinum resistor (Pt 1000), which is in good thermal contact with the Cu block. The sample is illuminated by a 450 W xenon lamp. The light beam is chopped, spectrally dispersed in a monochromator with a wavelength resolution of 2 nm, and focussed on the sensitive detector area of 70 mm^2. The total light intensity incident on the photodiode was determined to be 0.2 mW at λ = 450 nm. The relative intensity of the lamp spectrum was measured by the photoacoustic spectroscopy (PAS) method /22/ using a photoacoustic cell which absorbs light independent of the wavelength.
The photodiode current is measured by an I-V converter circuit maintaining short circuit operation conditions. The converter output is connected to a lock-in amplifier which operates at the light chopping frequency of 15 cps. Wavelength adjustment and data acquisition are achieved by a computer in connection with a stepper motor controller and a digital recorder via a general-purpose interface bus (GPIB).

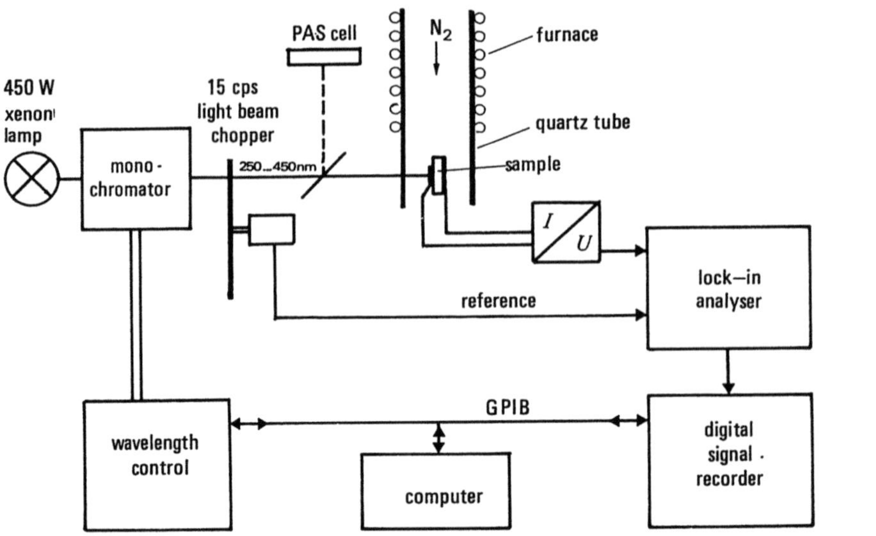

Fig. 17: Block diagram of experimental setup for UV–SiC–photodiode quantum efficiency measurements [22]

Figure 18 shows the quantum efficiency of a typical large-area SiC-photodiode versus the incident light wavelength in the temperature range from 295 to 673 K. Value and position of the maximum quantum efficiency varies only slightly within the temperature range investigated. The long wavelength cut-off at 400 to 450 nm corresponds to the intrinsic absorption edge of 6H-SiC. Caused by the temperature dependence of the band-gap energy - the band-gap decreases with temperature - the long wavelength cut-off shifts to longer wavelengths with increasing temperature of the photodiode. The short wavelength cut-off is governed by the steep increase of the absorption coefficient at shorter wavelengths in connection with the surface recombination velocity. The penetration depth of light with 250 nm wavelength can be estimated to be approx. 0.1 μm. For such wavelengths, the main contribution to the photocurrent originates from the implanted region which is close to the surface.

There, the carrier lifetime is strongly reduced by the high surface recombination velocity, which increases with increasing temperature. The maximum value of the quantum efficiency of photodiode No 31 - caused by superposition of the two limiting processes discussed before - is 37 % at λ = 315 nm and a temperature of 373 K. Also the position of the maximum quantum efficiency shifts with increasing temperature to longer wavelengths due to decreasing bandgap and increasing surface recombination velocity. The surface recombination velocity differs with surface quality and surface preparation (e.g. antireflection coating) and leads to a scatter of the location as well as of the absolute value of the maximum quantum efficiency. This fact is demonstrated for diode No. 6 in Fig. 19.

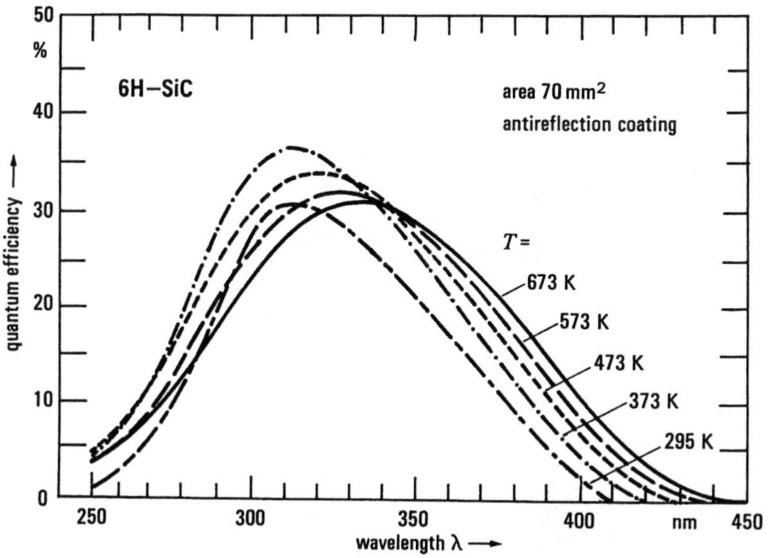

Fig. 18: Spectral quantum efficiency of a 6H SiC–UV photodiode at various temperatures, photodiode N° 31 [22]

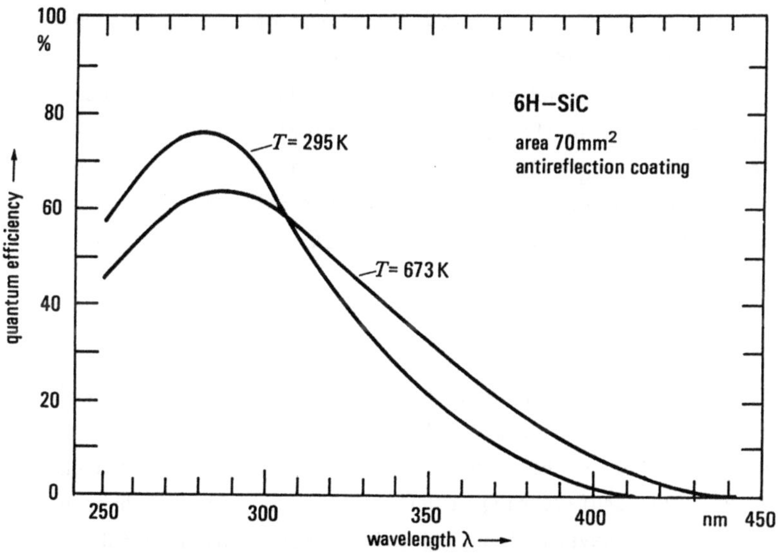

Fig. 19: Spectral quantum efficiency of a 6H SiCUV photodiode at various temperatures, photodiode N°6 [22]

This photodiode has apparently a nearly perfect surface leading to strongly lowered values of the surface recombination velocity. As a consequence, short wavelength cut-off and the maximum of the quantum efficiency shift to shorter wavelengths. The peak maximum quantum efficiency of this diode is located at λ = 280 nm and reaches the high value of 75 % at room temperature; the corresponding values at T = 673 K are λ = 290 nm and 63 %, respectively.

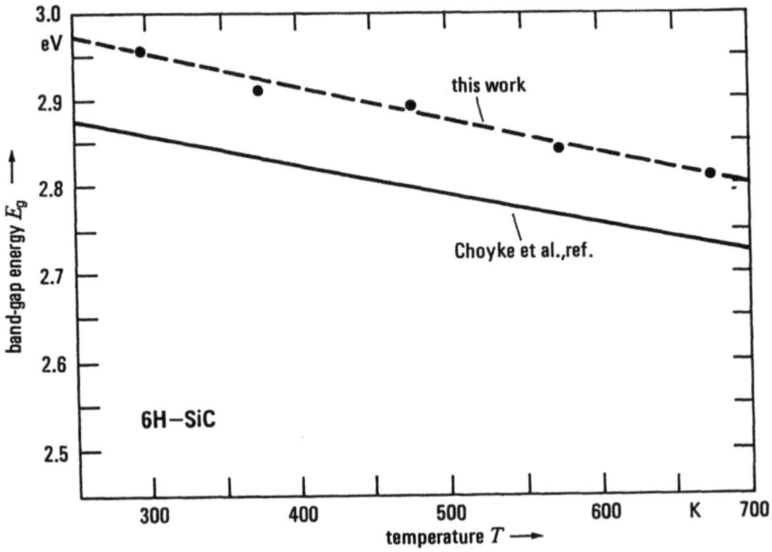

Fig. 20: Band–gap energy E_g of 6H–SiC versus temperature [22]

A detailed analysis of the shape of the long wavelength cut-off at various temperatures gives the band-gap energy and its temperature dependence. Our results are summarized in Fig. 20. We obtain a band-gap energy E_g = 2.95 eV at room temperature. This value is slightly greater than the values determined by W.J. Choyke et al. /24/ by similar experiments (E_G = 2.86 eV). The temperature coefficient of the band-gap energy E_G/ T is determined from the slope of the straight line in Fig. 20; it shows a linear dependence of the temperature $\Delta E_G/\Delta T = -3.8 \cdot 10^{-4}$ eV/K.

5. Conclusion

In the past few years, we have placed emphasis on the improvement of the crystalline quality of 6H-SiC by developing a modified Lely growth process, and improving the LPE-growth, and also on the transfer of modern processing techniques to fabricate SiC devices. The manufacturing technique of blue LED's has been perfected to the point where today several thousand chips a month could be fabricated. Ion implantation of N^{14} and rapid isothermal annealing (RIA) were employed to achieve n-type doping in epitaxial-grown 6H-SiC layers. Based on planar technique, UV photodiodes with shallow implanted pn-junctions have been developed.

References

1. H.J. Round: Electrical World 19 (1907) p. 309
2. J.A. Lely: Ber. dtsch. Keram. Ges. 32 (1955) p. 229

3. R.B. Campbell: IEEE Transactions on Industrial Electronics IE-29 (1982) p. 124
4. H. Matsunami, M. Ikeda, A. Suzuki, T. Tanaka: IEEE Transactions on Electr. Devices, ED-24 (1977) p. 958
5. H.J. van Daal: J. Phys. Chem. Solids 24 (1963) p. 109
6. W. v. Münch: J. of Electronic Materials 6 (1977) p. 449
7. M. Königer: Proceedings of Internat. Symp. on CdTe, Strasbourg (1971)
8. Yu.M. Tairov, I.I. Khlebnikov, V.F. Tsvetkov, M.A. Chernov: Soviet Phys. Crystallogr. 21 (1976) p. 234
9. Y.M. Tairov, Y.A. Vodakov: Electroluminescence, Springer Verlag (1977) p. 31
10. V.A. Dmitriev, Ya.V. Morozenko, I.V. Popov: Soviet Tech. Phys. Lett. (VIA) 12 (1986) p. 221
11. W.F. Kippenberg: Philips Research Reports 18 (1983) p. 161-274
12. Y.M. Tairov, V.F. Tsvetkov: Journ. of Crystal Growth 43 (1978) p. 209
13. G. Ziegler, P. Lanig, D. Theis, C. Weyrich: IEEE Transactions on Electron Devices ED-30 (1983) p. 277
14. L. Hoffmann, G. Ziegler, D. Theis, C. Weyrich: J. Appl. Physics 53 (1982) p. 6962
15. Y.M. Tairov, V.F. Tsvetkov: J. of Crystal Growth 52 (1981) p. 146
16. G. Pensl, W. Suttrop, H. Zhang: to be published
17. G. Pensl, R. Helbig, H. Zhang, G. Ziegler: Proceedings of the MRS Symposium, Anaheim (USA), April 21-23 (1987) p. 195
18. W. Suttrop, Diploma-Thesis, May 1987, Friedrich-Alexander-Universität Erlangen-Nürnberg
19. W. Suttrop, G. Pensl: to be published
20. R.W. Brander: Proc. IEE 116 (1969) p. 329
21. G. Ziegler, D. Theis: IEEE Transactions on Electron Devices ED-28 (1981) p. 425
22. P. Glasow, G. Ziegler, W. Suttrop, G. Pensl, R. Helbig: SPIE Vol 868 Optoelectronics Technologies for Remote Sensing from Space 1987
23. S.M. Sze: Physics of Semiconductor Devices (1981) Second Edition, p. 750
24. W.J. Choyke, L. Patrick: Proc. of the Conf. on SiC, Boston (1959), Pergamon Press (1960) p. 306
25. H. v. Seggern, private communication 1987
26. Th. Vetter, A. Winnacker, private communication 1987
27. Ch. Haberstroh, Diploma Thesis 1988, to be published
28. Landolt Börnstein New Series, Group III Vol. 17a, Springer-Verlag Berlin - Heidelberg - New York

VPE Growth of SiC on Step-Controlled Substrates

H. Matsunami[1], *K. Shibahara*[1], *N. Kuroda*[1], *W. Yoo*[1], *and S. Nishino*[2]

[1] Kyoto University, Yoshidahonmachi, Sakyo, Kyoto, Japan
[2] Kyoto Institute of Technology, Matsugasaki, Sakyo, Kyoto, Japan

Crystals of SiC were grown on (0001)Si faces of 6H-SiC at 1350-1500°C by CVD. Polytypes of the grown layers changed by introduction of off-orientation from (0001) faces. On natural surfaces of (0001) without off-orientation, twin crystals of 3C-SiC grew, whereas on off-oriented substrates single crystals of 6H-SiC grew. Off-orientation brings high density of atomic steps in the surface, which enables growth of 6H-SiC reproducibly even at 1400-1500°C. This temperature is 300-400°C lower than a typical growth temperature of around 1800°C for 6H-SiC. PN junction diodes of 6H-SiC fabricated at 1500°C showed very good rectifying characteristics. Blue light emission was observed in the forward bias region.

1. Introduction

The most popular polytypes of silicon carbide (SiC) are 3C- and 6H-types, which have energy gaps of 2.37eV and 3.08eV, respectively. In comparison with other wide-gap semiconductors such as GaN, ZnS, and ZnSe, SiC(both 3C- and 6H-types) has high electron mobility and it is much more stable. Thus, SiC is a promising material for high-temperature electronic devices. Particularly, 6H-SiC has a high breakdown field ($2\text{-}3.7\times10^6$ V/cm)[1] and high saturated electron drift velocity (2×10^7 cm/s)[2] which are attractive characteristics for high-power and high-frequency devices. Moreover, 6H-SiC has been investigated for the application to blue light-emitting diodes.

The lack of large crystals has prevented the practical use of SiC electronic devices. The sizes of 6H-SiC bulk crystals are currently 30mm in diameter and 13mm in height by the modified Lely method.[3] In the case of 3C-SiC, after our success in heteroepitaxial growth on Si,[4-6] wafers of 2 to 3 inches in diameter can be obtained. Recently, we reported that off-orientation of Si(100) substrates was effective in eliminating antiphase domains of 3C-SiC.[7,8]

In general, off-orientation substrates are effective to improve the crystal quality of epitaxially grown layers on them. Chemical vapor deposition (CVD) was adopted to grow both 3C- and 6H-SiC on 6H-SiC substrates by several researchers,[9-11] but no one has mentioned clearly the effect of off-orientation. Powell and Will reported epitaxial growth at low temperatures of 1320-1390°C onto the

Springer Proceedings in Physics, Vol. 34 Amorphous and Crystalline Silicon Carbide
Editors: G.L. Harris and C.Y.-W. Yang

crystal planes perpendicular to $(000\bar{1})$C faces.[12] However, usable areas of such planes are very small ($1.5x10mm^2$), and they mentioned nothing about electrical properties of the grown layers.

In this paper, we report that 6H-SiC can be homoepitaxially grown on off-oriented substrates prepared with (0001)faces of 6H-SiC even at 1400°C with a growth rate of 2 μm/h.[13,14] This temperature is much lower than the standard temperature of around 1800°C[15,16] ever known for the growth of 6H-SiC on (0001) faces. The growth mechanism is discussed using a model which deals with the density of atomic steps in the surface of the substrates. Characteristics of pn junction diodes prepared by this technique[14] are described.

2. Experimental

SiC was grown using a SiH_4-C_3H_8-H_2 system under atmospheric pressure in a horizontal reaction tube. Substrates were heated by rf induction. CVD growth was carried out at 1350-1500°C. Growth temperature was measured using an optical pyrometer by monitoring the temperature of a hole in a SiC-coated graphite susceptor.

Off-orientation substrates were prepared by angle-lapping of 6H-SiC single crystals. The plate-like 6H-SiC crystals with natural flat surfaces of (0001)Si or $(000\bar{1})$C were fixed on a specially devised jig with adhesive. The jig was designed to incline the normal direction of the wafer from the main crystal axes of [0001] or $[000\bar{1}]$ by a few degrees. Then, the crystals were lapped and polished with diamond paste. Polytypes of the grown layer were identified by the shape of etch pits, reflection high-energy electron diffraction (RHEED) pattern and photoluminescence (PL) spectra.

Undoped grown layers showed n-type conduction with a carrier concentration of $2x10^{16}$ cm^{-3} . To obtain p-type layers trimethyl-aluminum(TMA) or triethyl-aluminum(TEA) was introduced by a bubbling method using H_2 during the epitaxial growth. P-type conduction with a carrier concentration of $4\text{-}9x10^{17}cm^{-3}$ was obtained and measured.

3. Results and discussion

3.1 Crystal growth

A typical mosaic pattern was observed in the grown layer on a natural surface, where domains were separated by step-like or groove-like boundaries. The layer with such mosaic patterns was thought to be twin-crystalline(double positioning) 3C-SiC by observation of triangular-shaped etch-pits. This layer was identified as twin(double positioning) crystals of 3C-SiC by RHEED. This boundary was twin-boundary. Although growth conditions were widely varied to improve crystal quality, 3C-SiC with a single domain could not be obtained on natural surfaces. On the off-orientation substrates of a few degrees, epitaxial layers of very smooth surface without any boundaries were obtained. Etch pit observation using molten KOH etching technique could not give any information on polytypes, since the shapes were round, though different from those observed in 3C-SiC. The grown layers were identified as single crystalline 6H-SiC by the analysis of the RHEED patterns. Single crystals of 6H-SiC were grown in the temperature range from 1400°C to 1500°C. These temperatures are very low compared to a typical temperature of around 1800°C for the growth of 6H-SiC. The PL spectra could clearly classify the polytype of the grown layers.

Figure 1 shows the relationship between polytypes of the grown layers and the direction-and-magnitude of the off-orientation. When the off-angle was larger than 1.5°, 6H-SiC grew homoepitaxially. If the off-angle is smaller than 1°, twin 3C-SiC grows even on off-orientation substrates. These results are explained in relation to atomic steps in the surface which was introduced by off-orientation. Fundamentally, epitaxial growth occurs inheriting arrangements of atoms nearby the surface of substrates. When the natural surface is used, the density of atomic steps is very low, and the growth may occur on the flat surface rather than at the atomic steps. In this case, arrived atoms from the vapor phase inherit the stacking information of the atoms located just under the surface.

The natural surface has a very low off-angle from (0001), and there are low density of atomic steps in a macroscopic view point. Hence, the natural surface of 6H-SiC(0001) can be divided into two parts. In one part, the surface is occupied by the atoms belonging to the order of ABC which is the former half of the stacking order of ABCACB in 6H-SiC. In the other part, atoms belong to the latter part of ACB. On the ABC part 3C-SiC grows in accordance with the stacking order of ABC, and on the other part it grows with the order of ACB as shown in Fig.2(a), since the crystal grows inheriting the stacking order of the substrate. Thus, twin crystals of 3C-SiC grow on the natural surfaces of 6H-SiC.

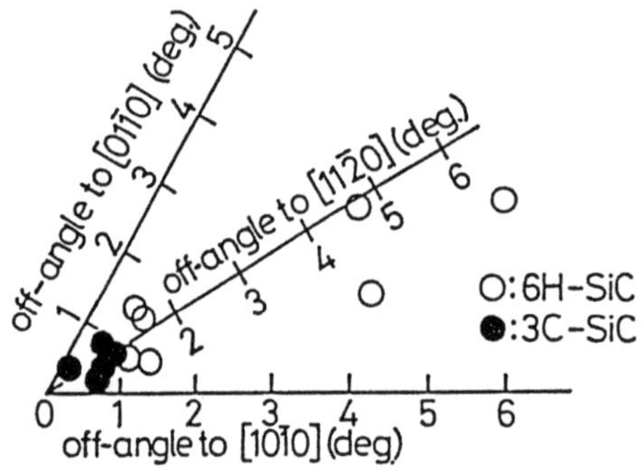

Fig.1 The relationship between polytypes of the grown layers and the direction-and-magnitude of off-orientation.

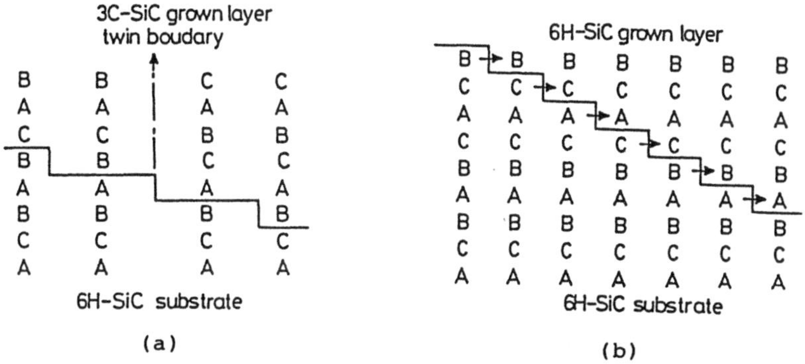

Fig.2 Model of (a) growth of twin crystalline 3C-SiC on 6H-SiC(0001) substrate without off-orientation, and (b) homoepitaxial growth of 6H-SiC single crystal on the substrate with off-orientation.

As an off-angle becomes larger, the density of the atomic steps increases. Then, arrived atoms onto the surface easily migrate to the atomic steps, since the distance of the atomic steps is very short. In this case, epitaxial growth occurs inheriting the stacking information around the atomic steps as shown in Fig.2(b). Thus, 6H-SiC layers were obtained easily on off-oriented substrates even at low temperatures. In the previous works[1,15,16] 6H-SiC grew at around 1800°C on a natural surface of 6H-SiC. At such a high temperature, surface migration becomes active, and atoms can reach to the atomic steps even if the density of the atomic steps is very low.

3.2 PN junction diodes

PN junction diodes with mesa-structure were fabricated. An undoped n-layer and Al-doped p-layer were consecutively grown on off-oriented n-type 6H-SiC substrates at 1500°C. The thickness of both n- and p-layers was about 2 μm. Using reactive ion etching(RIE) technique with an Al mask and CF_4+O_2 gases at 10Pa, mesa-structure was formed. The junction area of the diodes was $1.96\times10^{-3}cm^2$. Then, SiO_2 film was grown for surface passivation by thermal oxidation at 1050°C. For n- and p-type ohmic contacts Ni and Al-Si, respectively, were evaporated and alloyed. Figure 3 shows current-voltage characteristics of a fabricated pn junction diode. A large breakdown voltage of 100V was obtained, which gave a breakdown electric field of 2.4×10^6V/cm using observed graded junction characteristics. A value of $2-3.7\times10^6$V/cm was reported by von Muench and Pfaffeneder for a pn junction fabricated at 1800°C by the CVD method.[1] Thus, devices fabricated at lower temperatures showed comparable characteristics to those fabricated at high temperatures.

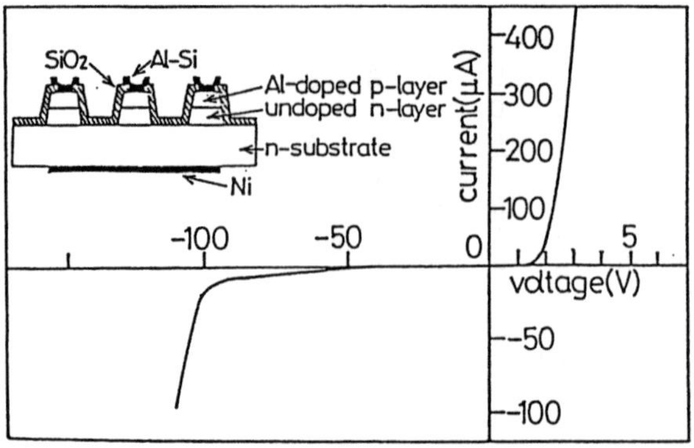

Fig.3 Current-voltage characteristics of a pn junction diode fabricated using homo-epitaxially grown layers of 6H-SiC.

These pn junction diodes showed blue light emission in the forward bias region. The spectra shown in Fig.4 agreed well with those of LEDs fabricated by liquid phase epitaxy(LPE).[17] The peaks are attributed to three origins: the highest-energy peak at 425nm (F peak) by free-exciton recombination, the peak at 455nm (E peak) by bound-exciton recombination at Al-related localized centers and other peaks (M peak) by donor-acceptor (N and Al) pair recombination.

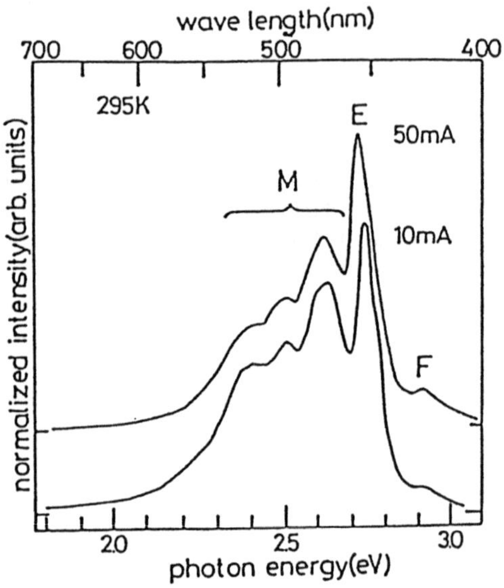

Fig.4. Electroluminescence spectra of a pn junction diode at room temperature.

4. Conclusion

Crystal growth of SiC was carried out on (0001)Si faces of 6H-SiC at 1350-1500°C by the CVD method. In this temperature range, usually 3C-SiC with twin boundaries grows on a natural surface. On off-oriented substrates, 6H-SiC grows even at 1400-1500°C, which is 300-400°C lower than a typical growth temperature. The change of polytypes by off-orientation of substrates was explained in relation to the density of atomic steps in the surface. Off-orientation seems to control the atomic steps. Fabricated pn junction diodes showed very good characteristics with a breakdown electric field of 2.4×10^6V/cm. The diodes emitted blue light under forward bias.

References

1. W.v.Muench and I.Pfaffeneder: J. Appl. Phys., **48** (1977) 4833.
2. W.v.Muench and E.Pettenpaul: J. Appl. Phys., **48** (1977) 4823.
3. K.Koga, T.Tanaka and T.Tanaka Ext. Abstr. 17th Conf. Solid State Devices and Materials (Business Center for Academic Societies, Japan, Tokyo, 1985) p.249.
4. H.Matsunami, S.Nishino and H.Ono: IEEE Trans. Electron Devices, **ED-28** (1981) 1235.
5. S.Nishino, H.Suhara and H.Matsunami: Ext. Abstr. 16th Conf. Solid State Devices and Materials (Business Center for Academic Societies, Japan, Tokyo, 1983) p.317.
6. K.Shibahara, S.Nishino and H.Matsunami: J. Cryst. Growth, **78** (1986) 538.
7. K.Shibahara, T.Saito, S.Nishino and H.Matsunami: Ext. Abstr. 18th Int. Conf. Solid State Devices and Materials (Business Center for Academic Societies, Japan, Tokyo, 1986) p.717.
8. K.Shibahara, S.Nishino and H.Matsunami: Appl. Phys. Lett., **50** (1987) 1888.
9. V.J.Jennings, A.Sommer and H.C.Chang: J. Electrochem. Soc., **113** (1966) 728.
10. S.Nishino, H.Matsunami and T.Tanaka: J. Cryst. Growth, **45** (1978) 144.
11. H.S.Kong, J.T.Glass and R.F.Davis: Appl. Phys. Lett., **49** (1986) 1074.
12. J.A.Powell and H.A.Will: J. Appl. Phys., **44** (1973) 5177.
13. N.Kuroda, K.Shibahara, W.Yoo, S.Nishino and H.Matsunami: Ext. Abstr. 19th Conf. Solid State Devices and Materials (Business Center for Academic Societies, Japan, Tokyo, 1987) p.227.
14. K.Shibahara, N.Kuroda, S.Nishino and H.Matsunami: Jpn. J. Appl. Phys.,**26** (1987).
15. W.v.Muench, W.Kuerzinger and I. Pfaffeneder: Solid-State Electron., **19** (1976) 871.
16. S.Nishino, A.Ibaraki, H.Matsunami and T.Tanaka: Jpn. J. Appl. Phys., **19** (1981) L353.
17. M.Ikeda, T.Hayakawa, S.Yamagiwa, H.Matsunami and T.Tanaka: J. Appl. Phys., **50** (1979) 8215.

Growth of β-SiC Heteroepitaxial Films on Vicinal (001) Si Substrates

L.G. Matus and J.A. Powell

NASA Lewis Research Center, Cleveland, Ohio 44135, USA

Heteroepitaxial growth of β-SiC films using vicinal (001) Si substrates has been studied. All SiC films that are grown by chemical vapor deposition using on-axis (001) Si substrates exhibit a high density of interfacial twinning, stacking faults, and antiphase disorder. Use of misoriented (001) Si substrates, tilted 1° to 4° toward the (110) plane, eliminates the antiphase disorder and improves the surface morphology of the films. Silicon substrates tilted toward the (100) plane were not effective in eliminating antiphase disorder. Observations of films grown on curved substrates indicate that tilt angles of 0.5^{o} may be sufficient to eliminate antiphase disorder. With respect to surface morphology, the on-axis films are less sensitive to the Si/C ratio of the input process gases than the misoriented films. The surface roughness for the on-axis films were nominally three times greater than the misoriented films.

1. Introduction

The NASA Lewis Research Center currently has a program to develop a new family of electronic devices and sensors based on silicon carbide (SiC) for aeropropulsion, space and earth-based applications. Applications include high temperature electronics for use in control and condition monitoring of aeropropulsion systems, high temperature integrated sensors, power electronics for space systems, and devices for communications purposes. As a semiconductor material, SiC can potentially meet several of NASA's needs for high temperature, high power, and high frequency electronic devices.

The current NASA Lewis SiC program, initiated in 1980, is based on the successful work of NISHINO et al. [1] who grew large area single crystal β-SiC heteroepitaxially on Si substrates. Since then, hundreds of SiC crystal growth runs have been completed resulting in a significant advancement in our understanding of the crystal growth process. Prior publications [2, 3] document this earlier work.

Despite continuing progress in the crystal growth, β-SiC films still contain many lattice defects. In particular, twins, stacking faults and antiphase boundaries (APBs) have been reported [4-7]. APBs occur as a common defect when a polar film, SiC in this case, is heteroepitaxially grown on a non-polar substrate [8]. To eliminate this particular defect in β-SiC films, Si substrates misoriented from the (001) plane have been used by SHIBAHARA et al. [9] and POWELL et al. [3]. This approach had been observed to eliminate APBs in GaP [10] and GaAs [11] heteroepitaxial films also grown on Si substrates.

The purpose of this paper is to present recent observations as a result of continuing to grow β-SiC films on vicinal Si substrates.

2. Experimental Procedures

The chemical vapor deposition growth system and the process for successful heteroepitaxial growth of β-SiC films on Si substrates are described in another paper [12]. The purposes of the growth experiments conducted in this series were threefold: (a) to further investigate the effectiveness of misoriented (001) Si

Springer Proceedings in Physics, Vol. 34 Amorphous and Crystalline Silicon Carbide
Editors: G.L. Harris and C.Y.-W. Yang

substrates in eliminating APBs, (b) to monitor the difference (if any) of the surface morphology of the β-SiC films grown on (001) misoriented Si substrates, and (c) to examine the effect on surface morphology when varying the Si/C atomic gas ratio of the input process gases.

When checking a β-SiC film for the existence of APBs, a decoration technique using in situ diborane gas was used [3]. This decoration method had been discovered during an earlier attempt to produce p-type β-SiC on n-type β-SiC using diborane as the doping gas. Once decorated, the APBs are readily observed with an optical microscope.

The Si substrates were cut from commercially polished, Czochralski grown, 75mm diameter wafers. Wafers were specified from the vendors with the following orientations: (001) Si, misoriented (001) Si tilted 1^0, 2^0, and 4^0 toward the (110) plane, and misoriented (001) Si tilted 1^0 and 4^0 toward the (100) plane. The wafers were p-type (boron doped) with resistivities in the range 5-25Ω-cm. To permit direct comparisons, multiple substrates (1 cm x 1 cm in size) with different orientations were used in the same growth run.

A stylus type profilometer was used to measure the surface roughness of the β-SiC films. Surface roughness was quantified by calculating the root-mean-square (rms) roughness value as follows: (1) a profilometer trace containing at least twenty peaks was obtained; (2) from the trace, the average peak-to-peak height was calculated in Å, and (3) the average peak-to-peak height was divided by $2\sqrt{2}$.

3. Results and Discussions

The effectiveness of using misoriented (001) Si substrates to eliminate antiphase disorder in heteroepitaxial β-SiC films was investigated by growing films on a variety of misoriented substrates. All on-axis substrates produced β-SiC films that exhibited APBs. The size of the antiphase domains was found to increase as the thickness of the film increased. This feature is shown in the surface micrographs of Fig. 1. A similar observation was reported by SHIBAHARA et al. [4]. It is apparent that annihilation of APBs occurs as the film thickness increases. In a previous publication [3], the use of the 2^0 misoriented Si substrates tilted toward the (110) plane was observed to be completely effective in eliminating antiphase disorder. To further investigate the applicability of misoriented substrates, additional growth experiments were performed using substrates misoriented 1^0 and 4^0 tilted toward the (110) plane and misoriented 1^0

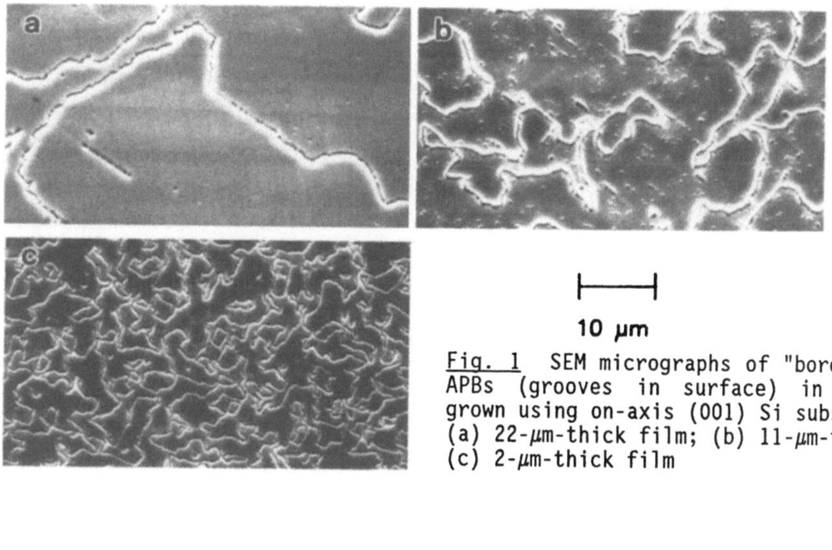

Fig. 1 SEM micrographs of "boron decorated" APBs (grooves in surface) in β-SiC films grown using on-axis (001) Si substrates: (a) 22-μm-thick film; (b) 11-μm-thick film; (c) 2-μm-thick film

and 4° substrates tilted toward the (100) plane. Again, substrates tilted toward the (110) plane were found to eliminate APBs. Substrates tilted toward the (100) plane, however, were not successful in preventing APB formation. This dependence on tilt direction was previously reported in GaAs films [13, 14] and was also observed by SHIBAHARA et al. for β-SiC films [15]. Although misoriented substrates have been successful in eliminating APBs, all samples continue to have a high density of stacking faults.

A comment on the magnitude of the tilt angle necessary to eliminate APBs is possible based on an observed region of APBs in a film grown on a 2° misoriented substrate. The APBs appear in a narrow band near and parallel to an {110} edge of the substrate as shown in the surface micrograph in Fig. 2a. Existence of this narrow band of APBs is made possible by the <u>in situ</u> HCl etching step that is performed as a substrate surface preparation step prior to β-SiC film deposition. This etching step causes a rounding of the Si surface near the edge of the substrate. This substrate rounding is schematically shown in Fig. 2b. Due to the curvature of the substrate, there is in effect a portion of the substrate surface that is on-axis. That is, a region exists where the surface is parallel to the lattice atomic planes of the misoriented substrate. Based on profilometer traces of the curved substrates, tilt angles of 0.5^{o} appear sufficient to eliminate antiphase disorder.

Surface morphology of the β-SiC films was monitored as a function of the Si/C atomic gas ratio of the input gases as well as a function of the tilt angle of the vicinal (001) Si substrates. Typical surface morphology features are shown in the Nomarski micrographs of Fig. 3, where β-SiC films grown using on-axis substrates are compared with films grown on misoriented substrates tilted 2° toward the (110) plane. A Si/C ratio of 0.42 represents our normal operating gas ratio and results in the best surface morphology. At a lower Si/C ratio (0.30), simultaneous deposition of large SiC particles occurs which are a hindrance when photolithographical processing is attempted. Higher Si/C ratios (0.56) produce surfaces that appear hazy to the unaided eye. Viewing these surfaces with Nomarski illumination shows a cross-hatched surface pattern for the on-axis grown film and a one-directional surface texture for the misoriented grown film.

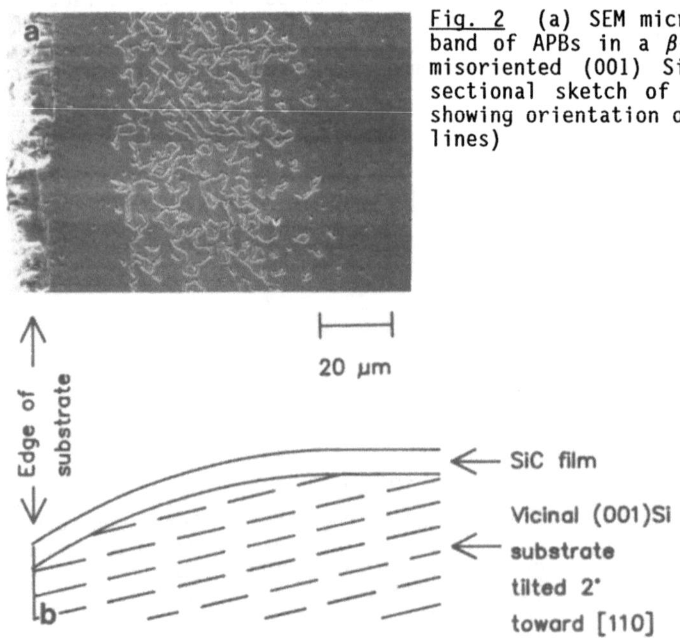

<u>Fig. 2</u> (a) SEM micrograph showing a narrow band of APBs in a β-SiC film grown on a 2° misoriented (001) Si substrate; (b) cross-sectional sketch of β-SiC film/Si substrate showing orientation of lattice planes (dashed lines)

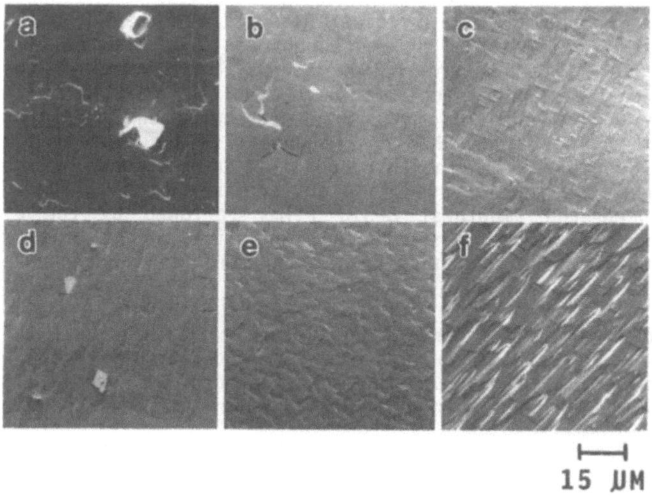

Fig. 3 SEM micrographs showing surface morphology of β-SiC films; (a), (b), and (c) films grown on on-axis (001) Si substrates at Si/C atomic input-gas ratios of 0.30, 0.42, and 0.56, respectively; (d), (e), and (f) films grown on misoriented (001) Si substrates tilted 2^0 toward the (110) plane at Si/C ratios of 0.30, 0.42, and 0.56, respectively

To quantitatively compare the surface roughness of β-SiC films, the rms surface roughness was measured for films grown on the various vicinal (001) Si substrates and at the three different Si/C atomic gas ratios. The data are plotted in Fig. 4. Three points can be concluded from these data. First, growing at a Si/C ratio

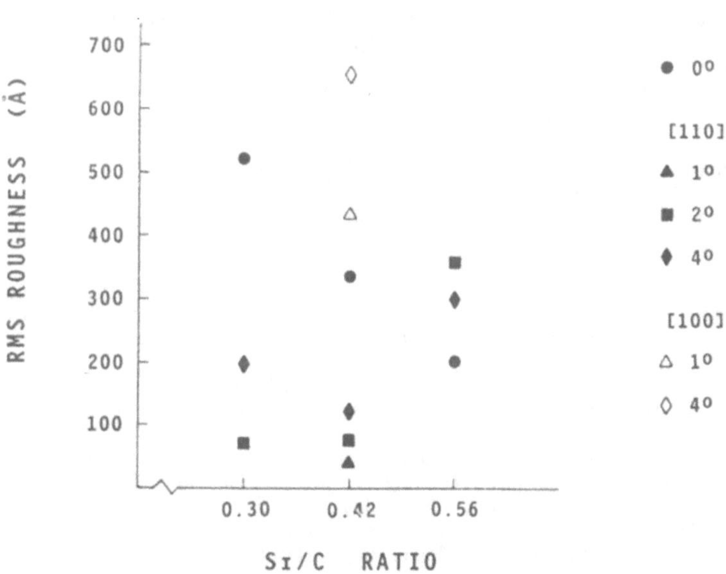

Fig. 4 RMS surface roughness for β-SiC films grown on various vicinal (001) Si substrates at three different Si/C atomic input-gas ratios

of 0.42 gives the best surface morphology for a variety of substrates based on the rms surface roughness values. Specifically, the rms roughness value for the film on a 1° misoriented substrate tilted toward (110) was 60 Å. Secondly, the misoriented substrates tilted toward (110) collectively resulted in films with very good surface morphology. This was not the case when growing on substrates tilted toward the (100) plane. These films had rms roughness values four to six times greater than the (110) counterparts. The third point to be mentioned is the order of the substrates responsible for improved film surface morphology. The 1° misoriented substrates produced the smoothest β-SiC films followed by the 2° misoriented substrates, the 4° misoriented substrates, and finally the on-axis (0°) substrates. Generally the surface roughness for the on-axis films were three times greater than the misoriented films.

It deserves to be mentioned as a final comment on surface morphology that the films grown using on-axis Si substrates were less sensitive to varying the Si/C atomic gas ratio than the misoriented films. The misoriented films especially decreased in surface quality when higher than normal Si/C atomic gas ratios were used.

4. Conclusions

Misoriented (001) Si substrates have been used to grow heteroepitaxial β-SiC films. APBs are eliminated when misoriented substrates tilted toward the (110) plane are used. Substrates tilted toward the (100) plane were not effective in eliminating APBs. An added benefit of growing on a misoriented substrate tilted toward the (110) plane is significantly smoother films. Although the film morphology is improved, it is sensitive to the Si/C atomic gas ratio.

References

1. S. Nishino, J. A. Powell, and H. A. Will: Appl. Phys. Lett. 42, 460 (1983)
2. J. A. Powell, L. G. Matus, M. A. Kuczmarski: J. Electrochem. Soc. 134, 1558 (1987)
3. J. A. Powell, L. G. Matus, M. A. Kuczmarski, C. M. Chorey, T. T. Cheng, and P. Pirouz: Appl. Phys. Lett. 51, 823 (1987)
4. K. Shibahara, S. Nishino, and H. Matsunami: J. Cryst. Growth 78, 538 (1986)
5. C. M. Chorey, P. Pirouz, J. A. Powell, and T. E. Mitchell: in Semiconductor-Based Heterojunctions: Interfacial Structure and Stability, edited by M. L. Green, J. E. E. Baglin, G. Y. Chin, H. W. Deckman, W. Mayo, and D. Narasinham (Metallurgical Society, Warrendale, PA, (1986) p. 115
6. S. R. Nutt, D. J. Smith, H. J. Kim, and R. F. Davis: Appl. Phys. Lett. 50, 203 (1987)
7. P. Pirouz, C. M. Chorey, and J. A. Powell: Appl. Phys. Lett. 50, 221 (1987)
8. H. Kroemer: J. Cryst. Growth 81, 193 (1987)
9. K. Shibahara, T. Saito, S. Nishino, and H. Matsunami: Extended Abstracts of the 18th International Conference on Solid State Devices and Materials, Tokyo, 1986, p. 717
10. M. M. Al-Jassim, J. M. Olson, and K. M. Jones: in Materials Problem Solving with the Transmission Electron Microscope, MRS Symposia Proceedings, Vol. 62, edited by L. W. Hobbs, K. W. Westmacott and D. B. Williams (Materials Research Society, Pittsburgh, PA, 1986), p. 49
11. M. Kawabe and T. Ueda: Jpn. J. Appl. Phys. 25, L285 (1986)
12. J. A. Powell and L. G. Matus: "Recent Developments in SiC (USA)", these Proceedings
13. S. Sakai, T. Soga, M. Takeyasu, and M. Umeno: in Heteroepitaxy on Silicon, MRS Symposia Proceedings, Vol. 67, edited by J. C. C. Fan and J. M. Poate (Materials Research Society, Pittsburgh, PA, 1986), p. 15
14. M. Akiyama, Y. Kawarda, S. Nishi, T. Ueda, and K. Kaminishi: ibid, p. 53
15. K. Shibahara, S. Nishino, H. Matsunami: Appl. Phys. Lett. 50, 1988 (1987)

Epitaxial Growth of 3C-SiC on a Si Substrate Using Methyltrichlorosilane

S. Nishino and J. Saraie

Department of Electronics and Information Science,
Faculty of Engineering and Design, Kyoto Institute of Technology,
Matsugasaki, Sakyo-ku, Kyoto 606, Japan

Single crystals of cubic SiC were heteroepitaxially grown by chemical vapor deposition(CVD) using methyltrichlorosilane(CH_3SiCl_3). The crystallinity of the grown layers was influenced by the substrate temperature and flow rate of CH_3SiCl_3. The cubic SiC(111) plane was grown on Si(111) substrates.

1. Introduction

Cubic SiC(3C-SiC) is an interesting material for electronic and optical device applications because of its high temperature stability and large energy gap. 3C-SiC has a bandgap energy of 2.35 eV at room temperature,[1)] an electron mobility of 1000 cm^2/V sec,[2)] and a drift velocity of $2.7x10^7$ cm/sec(at $2x10^5$ V/cm)[3)]. Furthermore, n-type and p-type conductions are easily obtained. These properties are particularly attractive for minority carrier devices, as well as majority carrier device applications. Epitaxial growth of 3C-SiC on foreign substrates has been reported by many investigators using two source materials, one for Si and the other for C. Recently, reproducible epitaxial growth of 3C-SiC on Si has been established by employing a buffer layer and subsequent CVD using SiH_4 and C_3H_8.[4,5,6)]

From the point of view of a simple handling CVD system, one source material containing both Si and C is desirable. Rai-Choudhury et al.[7)] reported the growth of polycrystalline 3C-SiC on Si substrates by CVD using dimethyldichlorosilane$(CH_3)_2SiCl_2$. They pointed out that unreacted Si was produced when the H_2 carrier gas was used, but it was not found when inert Ar gas was used. Jacobson[8)] reported that the highly oriented 3C-SiC was deposited on Si substrates at 1365 °C using the $(CH_3)_2SiCl_3$-H_2 CVD system. Methyltrichlorosilane(MTS; hereafter)is an another source material containing Si and C. Straughan et al.[9)] reported polycrystalline formation on the carbon tube by CVD using MTS. Bartlett et al.[10)] obtained single crystals of 3C-SiC on 3C-SiC(solution grown) substrates at 1600 °C. Knippenberg et al.[11)] reported SiC deposition on graphite rods in a temperature range 1227-2227 °C. They discussed the thermodynamic analysis and morphology of SiC films. The use of MTS for low temperature epitaxial growth of 3C-SiC on Si substrates has not been well understood. In this report, we describe the epitaxial growth of 3C-SiC on Si substrates at 1300 °C using MTS.

2. Experiment

Deposition of 3C-SiC was carried out by CVD using MTS. Vapor pressure of MTS was maintained at 150 Torr by keeping the bubbler at 20 °C. The reaction tube and temperature profile are shown schematically in Fig.1. The flow rates of the gases were controlled by mass flow controllers (Nippon Tyran Corp.). The reaction tube was ceramic, measuring 30 mm ID and 500 mm long. An SiC-coated, graphite cylinder was inserted in the ceramic tube as a liner tube to prevent the reaction of MTS with the ceramic tube. The susceptor was made of SiC coated graphite. The substrate temperature was measured by a Pt-Rh thermocouple connected in the susceptor. The substrate was then heated by an electric resistance heater. The reaction system has a gate valve and a loading chamber. The gate valve is necessary to prevent the reaction tube being exposed to the air when loading the sample. Consequently, many growth runs are possible without being interrupted by furnace heating. After the Si substrates were loaded into the loading chamber, the chamber was evacuated by a rotary pump and backfilled with H_2. Single-crystal growth of 3C-SiC consists of two steps: etching and crystal growth. The temperature program for epitaxial growth is shown in Fig.2. During this step, the H_2 carrier gas flow rate was maintained at 1000 sccm. After the gate valve was opened, the Si substrate was transferred to the etching position in the reaction tube. First, the Si substrate was etched at a temperature

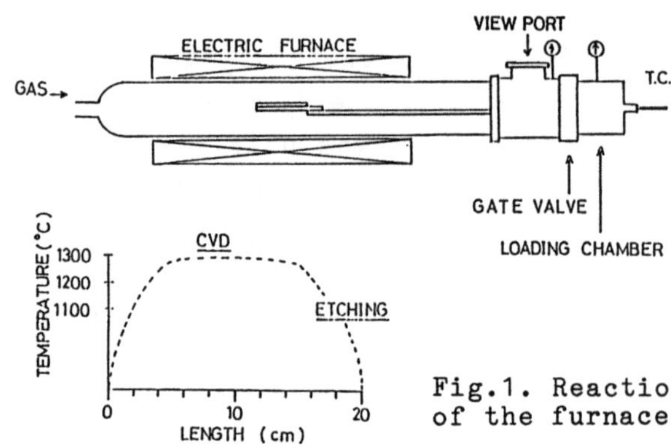

Fig.1. Reaction tube and temperature profile of the furnace.

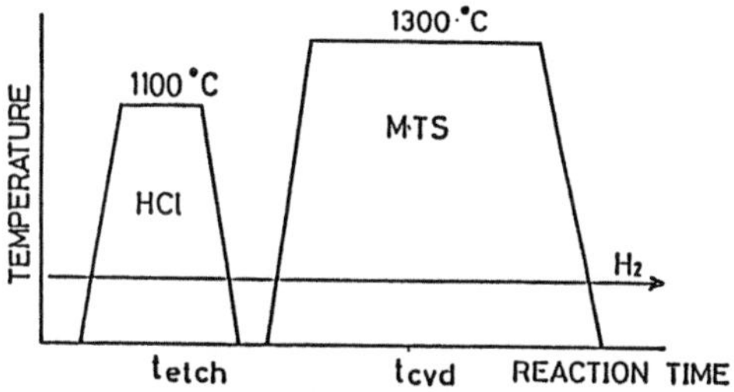

Fig.2. Temperature program of SiC single-crystal growth: t_{etch} and t_{cvd} mean the time for etching with HCl and single-crystal growth, respectively.

of 1100 °C in a HCl flow (5 sccm) for about 5 min. After the HCl gas was shut off , the substrate was pulled to the view port to examine the etched surface. Then, the Si substrate was transferred to its CVD position, and the MTS flow was added. Crystal growth was carried out for 5-15 min at a substrate temperature range of 1100-1300 °C.

3. Results and Discussion

The growth temperature and gas flow rate are important factors which determine the crystallinity and surface morphology of the grown layer. Si substrates with (111) orientation were used to study the substrate temperature dependence of the crystallinity of the cubic SiC film. After growing, the layers were examined by RED (Reflection Electron Diffraction) using an acceleration voltage of 80 kV.

The effect of MTS flow rate on the crystallinity of the grown layer was examined at a substrate temperature of 1300 °C. The diffraction pattern of cubic SiC, prepared at an MTS flow rate of 6 sccm, showed Debye rings which indicate a polycrystalline structure, as shown in Fig.3(a). Spots were superimposed on faint Debye rings in the film prepared at an MTS flow rate of 4 sccm, as shown in Fig.3(b). At an MTS flow rate of 2 sccm, the Debye rings disappeared , and the major characteristic pattern consisted of strong spots and streaky lines as shown in Fig.3(c). The weak spots were extra spots which could not be indexed using the regular reciprocal lattice. The diffraction pattern changed regularly by varying the direction of the incident beam. Therefore, the grown layer seems to be almost a single crystal, not a columnar structure. At an MTS flow rate of 1 sccm, the diffraction pattern consists of strong spots as shown in Fig.3(d). Thus, good single crystals were obtained at an MTS flow rate of 1 sccm.

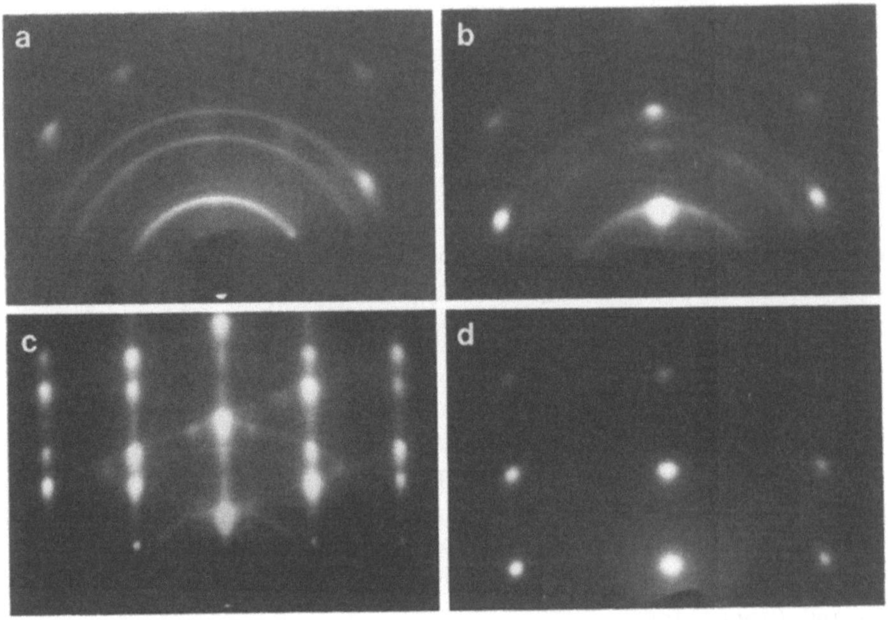

Fig.3. Reflection electron diffraction patterns for the film with various flow rates of MTS at a substrate temperature of 1300 °C: (a) 6 sccm,(b) 4 sccm,(c) 2 sccm, (d) 1 sccm.

By keeping the MTS flow rate at 1 sccm, the temperature dependence of crystallinity was examined. The diffraction pattern of cubic SiC prepared at 1100 °C and 1200 °C showed Debye rings, which indicate a polycrystalline structure, as shown in Fig.4(a,b). At a substrate temperature of 1300 °C, the diffraction pattern in Fig.4(c), consists of strong spots which are indexed, as shown in Fig.4(d). The growth rates of this film was about 200 Å/min. The epitaxial relationship between grown layers and Si substrates is as follows:

cubic SiC(111)//Si(111),

cubic SiC[110]//Si[110].

Surface morphology of the layers grown at three different temperature is shown in Fig.5. When the deposition was done at 1100 °C, the surface was covered with rounded grains with a diameter less than 1 µm. On the surface grown at 1200 °C, the size of the rounded grains became larger. When the film was deposited at 1300 °C, the surface was strongly influenced by the crystal habit of the substrate, and a triangular shape appeared.

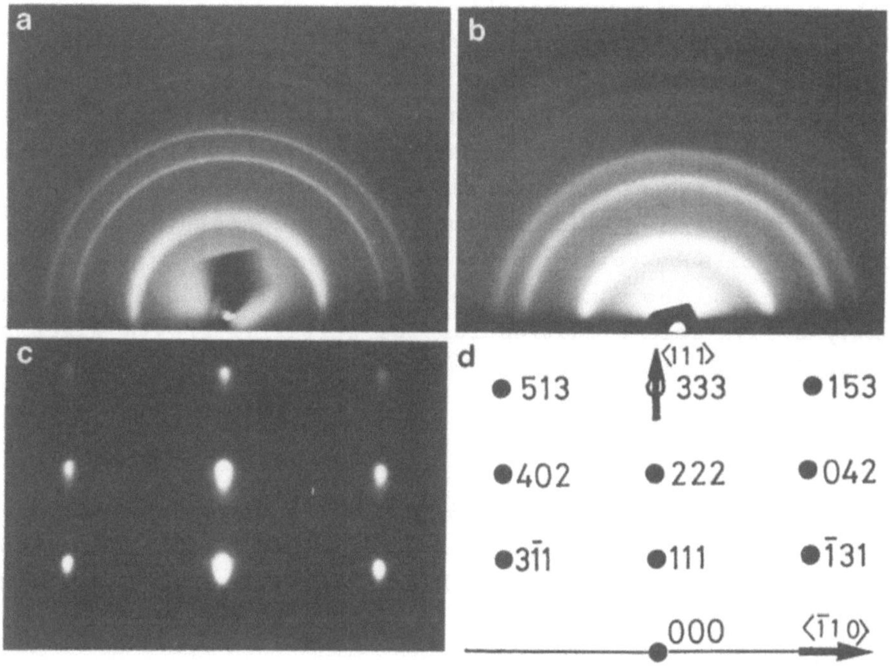

Fig.4. Reflection electron diffraction patterns for the film prepare with MTS flow rate of 1 sccm at different temperatures: (a)1100 °C,(b)1200 °C,(c)1300 °C,(d) indexed pattern of (c).

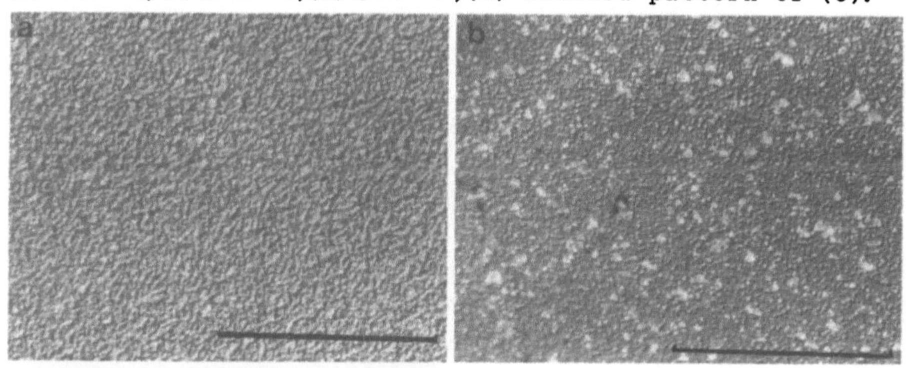

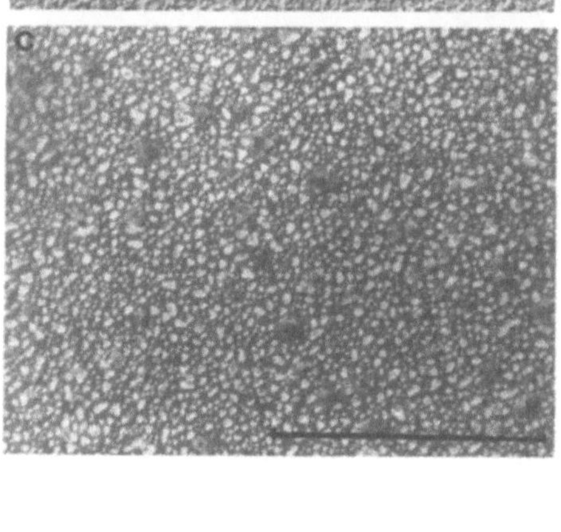

Fig.5 Nomarski microphotographs of the grown layer prepared at an MTS flow rate of 1 sccm at various substrate temperature:(a)1100 °C;(b)1200 °C;(c)1300 °C. Scale is 50 μm.

In the reproducible heteroepitaxial growth from SiH_4-C_3H_8-H_2 system, the buffer layer formation was a key factor in obtaining a single crystal.[5] Since single crystals were obtained without buffer layers in this experiment, different phenomena may happen at the initial stage of crystal growth. In our system, supplied gas (MTS) in the reaction tube is heated gradually along the flow direction. A large amount of HCl is formed at the high temperature region.[11] Thus, etching of Si substrates and the deposition of SiC might compete with each other at the initial stage of the crystal growth, and subsequently a favorable situation for the epitaxial growth might be established. The mechanism of crystal growth is still under study.

4. Conclusion

A single crystal of 3C-SiC on Si substrates was obtained at an MTS flow rate of 1 sccm and a substrate temperature of 1300 °C without buffer layer formation. The epitaxial relation was SiC(111)//Si(111) and [110]SiC//[110]Si.

References

1) D.S.Nedzvetskii, B.V.Novikov, N.K.Prokofeva, and M.B.Reifman, Sov.Phys.-Semicon.**2**,914(1969).

2) W.E.Nelson, F.A.Halden, and A.Rosengreen, J.Appl.Phys.**37**,33(1966).

3) D.K.Ferry, Phys.Rev. **B12**,2361(1975).

4) S.Nishino, J.A.Powell and H.Will, Appl.Phys.Lett. **42**,460(1983).

5) S.Nishino, H.Suhara, H.Ono, and H.Matsunami, J.Appl.Phys.**61**,4889(1987).

6) K.Shibahara, T.Saito,S.Nishino, and H.Matsunami, IEEE Electron Devices **EDL-7**,692(1986).

7) P.Rai-Choudhury and N.P.Formigoni; J.Electrochem.Soc.**116**,1440(1969).

8) K.A.Jacobson, J.Electrochem.Soc.**118**,1001(1971).

9) V.E.Straughan and E.F.Mayer;J.O'Connor and J.Smiltens ed., Silicon Carbide-A High Temperature Semiconductor (Pergamon Press,1960)p.94.

10) R.W.Bartlett and R.A.Muller; H.K.Henish and R.Roy ed.:Silicon Carbide -1968, Mat.Res.Bull.**4**,348(Pergamon Press 1969).

11) W.F.Knippenberg, G.Versupui and A.W.C. von Kemenade; R.C.Marshall, J.W.Faust,Jr. and C.E.Ryan ed.:Silicon Carbide-1973 (University of South Carolina Press,1974)p.92.

A Mechanism and Kinetics for Silicon Carbide Growth

M.E. Aluko

Department of Chemical Engineering and Materials Research Center,
Howard University, Washington, DC 20059, USA

A postulated mechanism for SiC growth based on adsorbed species SiH_2 and CH_4 is employed to obtain the intrinsic growth kinetics from SiH_4 and C_3H_8 in H_2 carrier gas as a function of gas partial pressures of the source species. Results of systematic experimental studies due to MINAGAWA and GATOS [1,2] provide much justification for the theoretical modeling work.

1. INTRODUCTION

The ability to grow uniform epitaxial films of semiconductor materials such as silicon carbide is of prime interest in the microelectronic industry where CVD has become increasingly popular. This desire is complicated by the fact that several processes and constraints simultaneously govern the CVD process - thermodynamic constraints, homogeneous/heterogeneous reactions and fluid-dynamic (mass and heat transfer) effects. Process analysis is simplified when one of these considerations dominates the overall behavior, which fact can often be deduced from experimental data.

While the exact nature of molecular-level crystal growth is not yet fully understood, it is well-accepted that adsorbed species are vital to crystal growth. Thus, deposition must be mediated by adsorbed species or molecules which chemisorb from the gas phase onto active sites on the growing surface, undergo some chemical transformation (preceded or followed by surface migration) before incorporation into the crystal structure. These facts provide some impetus to the study of growth kinetics based on postulated adsorbed species, as evidenced by several reports in connection with silicon growth [3-6] and gallium arsenide [7,8]. To the author's knowledge, no such work has been reported for SiC growth.

In this paper, we show that using SiH_2 and CH_4 as the ad-species, it is possible to explain the growth behavior of SiC from SiH_4-C_3H_8-H_2 mixtures based on a set of consistent experimental results of MINAGAWA and GATOS [1,2].

2. POSTULATED MECHANISM FOR SiC GROWTH

The overall reaction between SiH_4 and C_3H_8 to form SiC is as follows:

$$3SiH_4 + C_3H_8 \longrightarrow SiC(s) + 10\,H_2 \; . \qquad (1)$$

In the light of studies on SiH_4 in connection with silicon growth [9,10], thermodynamic calculations [11,12] and kinetics-sensitivity studies [13] on the SiH_4-C_3H_8-H_2 system, it is postulated here that the following sequence of steps occur:

$$SiH_4 \overset{K_1}{\longleftrightarrow} SiH_2 + H_2 \qquad (2)$$

$$SiH_2 + X \overset{K_2}{\longleftrightarrow} SiH_2.X \qquad (3)$$

$$C_3H_8 \overset{K_3}{\longleftrightarrow} CH_4 + C_2H_2 + H_2 \qquad (4)$$

$$CH_4 + X \overset{K_4}{\longleftrightarrow} CH_4.X \qquad (5)$$

$$SiH_2.X + CH_4.X \overset{k_R}{\longrightarrow} SiC(s) + 3H_2 + 2X \qquad (6)$$

Here X denotes the active substrate site, D.X denotes the species D in the adsorbed state, and SiC(s) is the silicon carbide incorporated into the crystal structure. The K_i's are equilibrium constants and k_R is the reaction rate constant for the bimolecular surface reaction Eqn. (6).

We now assume that Eqn. (6) is rate controlling (mass transfer is not limiting). Then following the Langmuir-Hinshelwood formulation [14] widely employed in heterogeneous catalysis, the "intrinsic" rate of reaction for growth per unit surface area, r, can be written as

$$r = k_R\,[SiH_2.X].[CH_4.X] \qquad (7)$$

where $[SiH_2.X]$, $[CH_4.X]$ are the surface concentrations of the respective adsorbed species. Assuming equilibrium for all the other reversible steps, almost complete conversion of C_3H_8 via Eqn. (4), and carrying out a balance on all active and occupied sites, the following growth equation is readily obtained from Eqns. (2)-(6) (G in microns/hour, A is a conversion factor from units of r to those of G):

$$G = \frac{Ak_R K_1 K_2 K_3^2 K_4 [K_3 - P_{H2} P_{C3H8}] P_{SiH4}}{[(P_{H2} + K_1 K_2 P_{SiH4})(K_3 - P_{H2} P_{C3H8}) + K_3^2 K_4]^2} \quad . \tag{8}$$

We now consider special simplifications of Eqn (8) as suggested by the experimental data of MINAGAWA and GATOS [2]:

(a) constant P_{C3H8} and P_{H2}, varying P_{SiH4} :

$$G = \frac{aP_{SiH4}}{[1 + bP_{SiH4}]^2} \tag{9}$$

(b) constant P_{SiH4} and P_{H2}, varying P_{C3H8} :

$$G = \frac{c - P_{C3H8}}{[d - eP_{C3H8}]^2} \tag{10}$$

(c) constant Si/C ratio; varying P_{SiH4}:

$$G = \frac{[f - gP_{SiH4}]P_{SiH4}}{[1 + hP_{SiH4} - mP^2_{SiH4}]^2} \tag{11}$$

where the parameters a,b,c, ...,m are all functions of the equilibrium constants K_i and relevant fixed partial pressures.

3. DISCUSSION OF RESULTS

Eqn. (9) shows that as a function of P_{SiH4}, G should exhibit an increase at small values of P_{SiH4}, attain a maximum at some intermediate value, then fall off to zero at yet higher partial pressure values. Eqn. (10) shows that as P_{C3H8} increases from near zero, G should increase from a near constant value to a high value since the denominator of Eqn. (10) decreases faster than its numerator. The trend predicted for G as a function of P_{SiH4} from Eqn. (11) is similar to that from Eqn. (10) except the increase at low partial pressure is not as pronounced, and the high-partial pressure dependence is more complex.

All of these predictions are in good agreement with experimental trends reported by MINAGAWA and GATOS [2] in their growth of α-SiC from the silane and propane vapor phase (H_2 carrier gas) in a vertical reactor at 1600°C (see Figures 1 a-c). Note that in Figure 1b, a change in C/Si for fixed P_{SiH4} corresponds to changing P_{C3H8}.

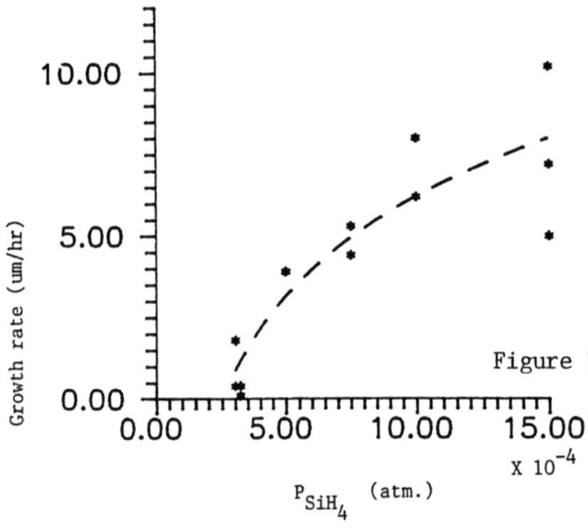

Figure 1a: Growth rate vs partial pressure of silane. C/Si = 1, T = 1600°C.

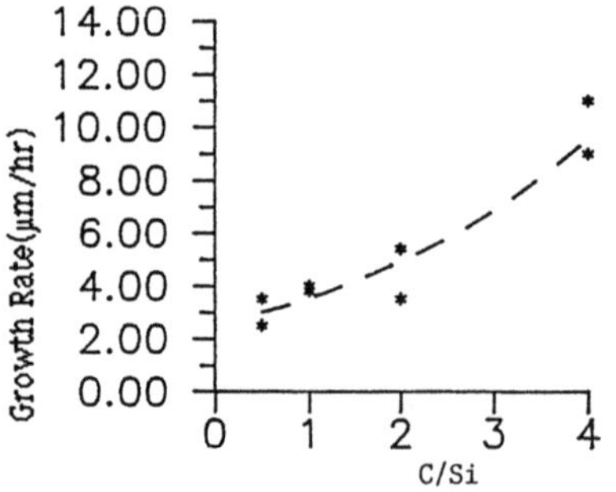

Figure 1b: Growth rate vs C/Si. $P_{SiH_4} = 5x10^{-4}$ atm, T = 1600°C

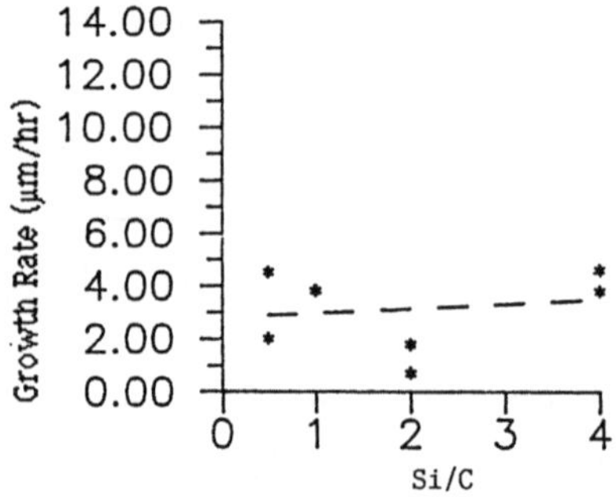

Figure 1c: Growth rate vs Si/C. $P_{C_3H_8} = 1.67x10^{-4}$ atm, 1600°C.

4. CONCLUSION

It has been shown that the growth of silicon carbide from SiH_4 and C_3H_8 in H_2 can be well described in terms of kinetics of the adsorbed species SiH_2 and CH_4. This does not in any way provide proof of their participation, but certainly indicates that they are possible candidates to be targeted for surface analytical investigation of ad-species. Finally, a wider range of operating conditions (for example different temperatures) and a similar study of β-SiC growth is desirable before any firm conclusions can be made.

REFERENCES

1. S. Minagawa, and H.C. Gatos, Jap. J. Appl. Phys. 19 (7) 844-849 (1971).
2. S. Minagawa and H.C. Gatos, Jap. J. Appl. Phys. 10 (12) 1680-1690 (1971).
3. W.A.P. Claassen and J. Bloem, J. Cryst. Growth 51 (1981) 443-452.
4. A.E.T. Kuiper, C.J.H. van den Brekel, J. de Groot and G.W. Veltkamp, J. Electrochem. Soc. 129 (1982) 2288-2291.
5. H.H. Lee, J. Cryst. Growth 69 (1984) 82-90.
6. E.C. Stassinos, T.J. Anderson and H.H. Lee, J. Cryst. Growth 73 (1985) 21-30.
7. J. Korec and M. Heyen, J. Cryst. Growth 60 (1982) 297-306.
8. V.N. Mani, R. Dhanasekaran and P. Ramasamy, Semicond. Sci. Technol. 2 (1987) 73-77.
9. W.A.P. Claassen and J. Bloem, Phillips J. Res. 36 (1981) 124.
10. C.G. Newman, H.E. O'Neal, M.A. Ring, F. Fleska and N. Shipley, Intern. J. Chem. Kinet. 11 (1979) 1167.
11. A.I. Kingon., L.J. Lutz, P. Liaw and R.F. Davis, J. Amer. Ceram. Soc. 66 (1983) 558-566.
12. M.E.Aluko, G.L. Harris, K.H. Jackson, and M.G. Spencer (submitted).
13. C.D.Stinespring and J. Wormhoudt, "Optimization of Silicon Carbide Production", NASA-Lewis Report ARI-RR-485, July 1985. 14. Hayward, D.O. and B.M.W. Trapnell, "Chemisorption" (Butterworths, London, 1964).

Mechanisms of Epitaxial Growth of ß-SiC on Silicon Substrate by Chemical Vapor Deposition and Nucleation of Defects

P. Pirouz[1], *T.T. Cheng*[1], *and J.A. Powell*[2]

[1] Department of Materials Science and Engineering, Case Western Reserve University, Cleveland, Ohio 44106, USA
[2] NASA Lewis Research Center, Cleveland, Ohio 44135, USA

In the growth of many compound semiconductors on a substrate by chemical vapor deposition (CVD or MOCVD), the one-step process, in which the substrate is heated to the growth temperature and then the gases are allowed to enter the reactor, results in a polycrystalline film. Recently, a two-step method has been used where the substrate is ramped from a low- to the final growth temperature in the presence of the reacting gases and then the deposition continued. The resulting film is then monocrystalline, although the defect density is rather high. The reason for the difference in the nature of growth in the one- and two-step processes is often attributed to the formation of a buffer layer in the latter case, although the nature of this layer has never really been explained clearly. In this paper, the classical mechanisms of nucleation and growth are used to explain the processes taking place in the one- and two-step methods and the resulting differences. Epitaxial growth of ß-SiC on (001) silicon substrate will be considered as an example. Mechanisms of formation of various defects commonly observed in the films grown in this way will then be discussed. The importance of lattice and thermal mismatch between the film and the substrate are considered and compared to the role of nucleation and growth in defect formation. Many examples of defects in epitaxially grown ß-SiC observed by conventional as well as high resolution transmission electron microscopy will be shown in light of the proposed mechanisms.

Springer Proceedings in Physics, Vol. 34 **Amorphous and Crystalline Silicon Carbide**
Editors: G.L. Harris and C.Y.-W. Yang

Part II

Growth of Amorphous Silicon Carbide

Present and Future Applications of Amorphous Silicon Carbide

S. Furukawa

Department of Applied Electronics, Graduate School of Science & Engineering, Tokyo Institute of Technology, 4259 Nagatsuda, Midori-ku, Yokohama 227, Japan

1. INTRODUCTION

Enormous progress in silicon devices is evident in every segment of industry and society. We expect the trend to continue toward ever-increasing levels of integration and speed in integrated circuits for signal processing as well as increasing performance in energy generation and control.

In conventional Si devices including ICs, however, there are many problems to be solved or points to be noted as follows:

1) The fine patterning in bipolar or Bi-CMOS IC for high density integration and for high speed will cause an increase of the base resistance of bipolar transistors. The increase of the resistance will degrade device performances at high frequency. If hetero-emitter junction bipolar transistors (HBT) can be fabricated by using Si-related materials and technology, this problem would be solved just as in compound semiconductor devices.
2) High power devices and special devices for use under harsh conditions (high temperature, etc.) will have to be used under worse and worse conditions in the future. Such devices need materials having an effectively wider band gap than Si.
3) In the future, an IC chip having sensors and processors integrated, that is, a smart sensor, will become very important. Such sensors should have high stability. In other words, the chemical stability of the films used in the IC should be much higher in comparison with those used in conventional Si technology.
4) In solar batteries, wide band gap materials are useful to realize high performance batteries as window materials for incident light or/and as a built-in potential barrier for carriers induced by the photoelectric effect.

It can be said that such problems would be solved by using wide band gap Si related materials such as crystalline silicon carbide and so on.
The scope of this paper is to show that amorphous (a-) and/or micro-crystalline (μc-) silicon carbide are also potentially capable of solving the above problems.

2. NEW APPROACHES OF Si TECHNOLOGY

In the past, solid state electron devices engineers have tried to design devices by considering physical constants of industrially available electronic materials, i.e., conventional materials, the typical example of which is silicon. In order

Springer Proceedings in Physics, Vol. 34 **Amorphous and Crystalline Silicon Carbide**
Editors: G.L. Harris and C.Y.-W. Yang

to solve the above problems mentioned in the introduction, we can say that it is basically necessary to create new electronic materials having various physical constants which are not found in such conventional ones.

Compound semiconductors, especially those composed of three or four elements, should be classified as newly created materials. In other words, such compound materials have only been synthesized during the past 20 years, just for use in heterojunction devices. We can change their band gap energies while retaining their lattice constants to realize double heterojunction lasers as well as HEMTs having good performances.

Table 1 shows examples 1-4 of material constants which the device designers want to change for specified applications in a, b, ... Detailed explanations should not be necessary. We should note that material constants shown in the table are average effective ones. That is, even if semiconductors having low dielectric constants were needed for the purpose of realizing low loading capacitances in IC, we could not realize such semiconductors by single-phase compound materials. The reason is that dielectric constants of covalent or quasi-covalent bonded materials are as high as, for example, 12 for Si and 14 for GaAs. However, we can get semiconductor materials having an effectively low dielectric constant by using a layered structure such as Si on insulator (SOI). As is well known, since most semiconductor devices are often fabricated near the surface of the semiconductor, we can substitute insulators having low dielectric constants for deeper regions of the semiconductor so as to decrease loading capacitance.

For this reason, we should have more interest in composite electronic materials rather than single phase ones, especially for the cases when we want to get more freedom in selecting material constants in the field of Si technology. This is because only restricted elements are allowed in Si technology.

Table 1. The material constants which it is desirable to be able to change for device applications and device performances.
BPT: Bipolar transistor, LD: Laser diode, RD: Rectifying diode, HEMT: High electron mobility transistor, SISFET: Semiconductor insulator semiconductor FET, OEIC: Opto-electronic IC, DRAM: Dynamic random access memory.

1. Band Gap E_G
- a. High Injection Efficiency in BPT
- b. Variation of Emitted Wavelength in LD
- c. High Breakdown Voltage in BPT, RD
- d. Carrier Confinements in LD, HEMT
- e. Epitaxial Gate Insulator in SISFET
- f. Radiation Resistant

2. Dielectric Constant ε_s
- a. Low Capacitive Loading in IC
- b. Wave Guide in OEIC
- c. High Capacitance in DRAM

3. Heat Conductivity κ
- a. High Power
- b. High Integration Density

4. Thermal Expansion Coefficient α
- a. High Power
- b. High Reliability
- c. Ultra-Microfabrication

3. MATERIALS AND PROCESSES NEEDED IN FUTURE DEVICE FABRICATION

As just mentioned, it can be said that the utilization of heterojunction and/or hetero-materials layered structures is, at least, necessary even in Si technology. There are many wider band gap materials than Si. However, from the viewpoint of Si process technology, it is better if such materials are composed of group IV elements, such as C, Ge and of Si-oxides, Si-nitrides, since other materials usually dope Si, so decreasing yield and reducing reproducibility of Si IC production.

For this reason, we should restrict the candidates of hetero-materials to Si to SiC, SiGe, SiO_2 and Si_3N_4. But, for example, crystalline SiC needs as high a temperature as about 1000 °C to be grown on a Si substrate. But, if we want to use such a SiC/Si structure for emitter/base region of high speed HBTs, the process temperature should be as low as about 600 °C. The reason is that the out-diffusion of base impurity into collector and emitter regions will cause a thicker base which degrades the high frequency performance of the HBT.

From the viewpoint of device mechanism, it is not necessary to use crystalline emitter for HBTs. The carrier flowing in the emitter is the majority one, so that a long lifetime of the carrier is not necessarily needed. A more important factor to be considered is the resistivity of the material as well as the ease of preparation of ohmic contacts.

Such consideration gives us a new concept to realize the composite electronic materials related to Si technology. We can utilize various combinations of crystalline phases as shown in Fig.1. By selecting materials and choosing combinations of crystalline phases, we could have new material constants which could not be utilized by easy fabrication processes such as low temperature, no high vacuum system, ... in the past.

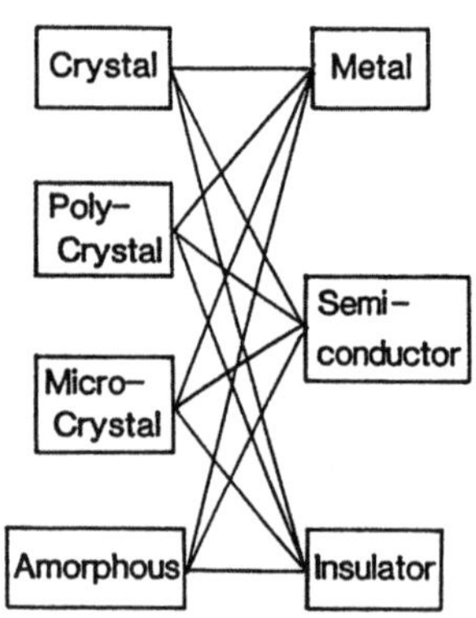

Fig. 1. Combinations of various materials and various crystalline phases.

4. WHY AND WHICH HETEROJUNCTION OF a-SiC/c-Si

4.1 Motivation to study a-SiC emitter HBTs

As an example of a new approach of Si technology, we shall explain why a heterojunction of a-SiC/c-Si is attractive for fabricating Si HBT's.

In order to permit an increasing level of integration in ICs, the power consumption of each device in the circuit should be decreased without decreasing the speed. Most IC designers seem to be of the opinion that this could be achieved by a circuit formed of bipolar and CMOS transistors, so called Bi-CMOS. (Recently complementary bipolar logic circuit -CBL circuit- became another candidate.) However, the fabrication process of a bipolar transistor is much more complicated than that of CMOS FETs. Therefore, a simpler process for building the emitter region will need to be developed. One method is a deposited emitter, which is not necessarily crystalline. Moreover, the cut off frequency of the bipolar transistors integrated should be high. This means that very thin base regions as well as low base resistance are needed. In order to decrease or to keep base resistance at a low value even for the case of a thin base region, a high impurity doping in the base is necessary. But this causes degradation of the current gain. To overcome this problem, HBT is very much appreciated, as mentioned in Section 1. In this case, a heterojunction emitter composed of material having a wider band gap or a base having a narrow band gap are alternative choices. Table 2 shows the combinations of emitter and base materials in HBT. We should remember that HBT is needed especially for the case of transistors with high cut-off frequency. That is, the base thickness must be thin. Therefore, a high temperature process cannot be used because it causes auto-diffusion of base impurity which broadens the base thickness [1],[2].

Such consideration leads us to the conclusion that amorphous or micro-crystalline materials having a wider band gap such as a-SiC should be used as emitters, since the deposition of the material can be achieved at relatively low temperatures (<600 °C).

Table 2. Combinations of emitter and base materials in Si-HBT.

Emitter	Base
Si	Si_xGe_{1-x}
SiC	Si
SIPOS	Si
GaP	Si
GaAs	Si
a-Si	Si
a-SiC	Si

4.2 Electrical Properties of a-SiC/c-Si Diode

There has been no report concerning electrical properties of a-SiC/c-Si. It is necessary, first of all, that the interface between the a-SiC and c-Si layered structure have good electrical characteristics so that the heterojunction can be used in HBT. Fig. 2 shows I-V characteristics of a p-type a-SiC/n-type c-Si diode, in which an a-SiC layer has been deposited by the L-coupled plasma CVD method. The n-value of the diode is found as good as 1.1 for the case of planar diode structure [3]. This means that localized state density at the interface is small in this system, for the following reasons. The number of dangling bonds at the interface tends to become small due to the flexibility of a-SiC networks so as to be easily bonded with Si atoms at the surface of crystalline substrate and furthermore the dangling bonds at the interface tend to be terminated by hydrogen atoms. Further evidence that the interface is electrically good can be seen from Fig. 3, which shows the performance of a solar battery made of a-SiC/c-Si structure [4]. A fairly good result was observed. From such experimental results, we conclude that an amorphous/crystalline semiconductor junction can be applicable to more complicated devices, such as HBTs.

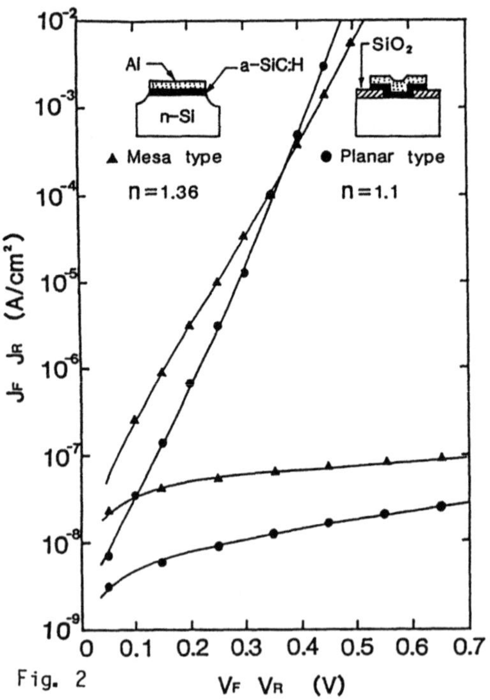

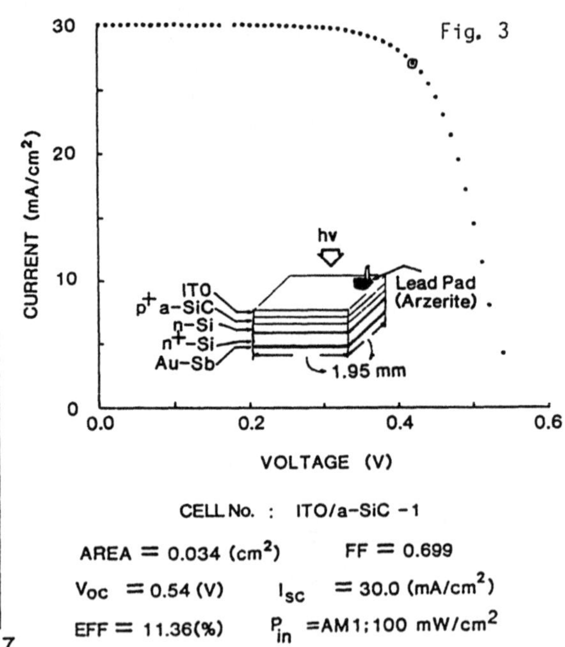

Fig. 2. I-V characteristics of p-type a-SiC/n-type c-Si heterojunction diodes.

Fig. 3. I-V characteristics of an a-SiC/c-Si structure solar battery.

4.3 Performance of a-SiC Emitter HBT

We have fabricated pnp a-SiC emitter HBTs, of which the current gain h_{FE} is shown in Fig. 4 as a function of collector current I_C [5]. Relatively high current gain has been observed. By observing the temperature dependence of the performance of this pnp transistor and pn diode, we have derived the band structure of this transistor as shown in Fig. 5(a). Since this band structure has only just been derived, there may be some error in it. However, we can at least conclude that the npn transistor is superior to the pnp one, the potential barrier for the majority carrier of the base is much higher in the case of npn transistors, as shown in Fig. 5(b). (Detailed discussion should be done by using Gummel number. See Ref. [6]) Fig. 6 shows I-V characteristics of an npn a-SiC emitter HBT, the maximum current gain of which is observed to be 180 [7]. From the above results, we can say that an a-SiC/c-Si structure is very useful for Si-HBT. But careful observation of the characteristics reveals that the emitter resistance due to high resistivity of a-SiC degrades the performance. Therefore, it will be necessary to decrease the specific resistance as well as to make a thin emitter region on which can be fabricated stable ohmic contacts. One of the methods to overcome this problem may be the use of micro-crystalline SiC instead of amorphous.

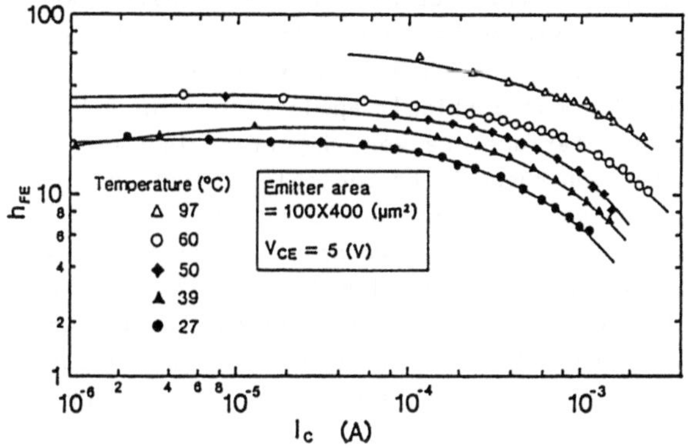

Fig. 4. The common emitter current gain, h_{FE}-collector current, I_C characteristics at various temperatures.

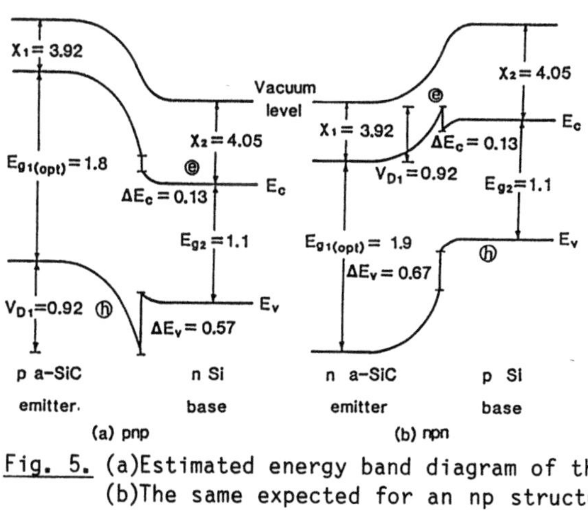

Fig. 5. (a)Estimated energy band diagram of the a-SiC/c-Si pn heterojunction, (b)The same expected for an np structure.

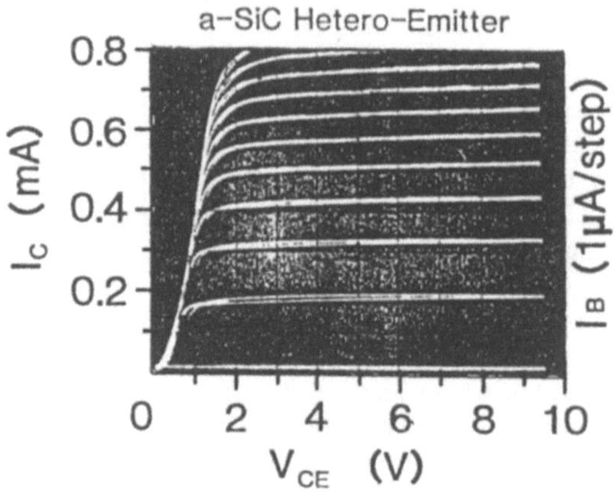

Fig. 6. I_C-V_{CE} characteristics of the npn a-SiC emitter transistor.

5. FUTURE AMORPHOUS SiC APPLICATIONS

In the preceding section, we have shown the feasibility of a-SiC emitter HBTs as an example. We should remember that such HBTs would be useful in microwave power transistors which usually need ballast resistances between emitter electrodes and emitter region. We can expect other applications of a-SiC, as shown in Table 3. In this table, solar cells and HBT are the most difficult applications for a-SiC, because the electronic properties at the interface are very important in these devices. But, still, such devices have been shown to be feasible. Other applications are also important. We can say such applications should be more earnestly considered. For such applications, we could develop a process to deposit a dense, chemically and thermally stable, and low resistive film at relatively low temperature. In Table 3, high temperature device application means that the device can be used at high temperature without change of the electronic properties. If one wants devices for extremely high temperatures, crystalline SiC or C (i.e. diamond) should be used as device materials.

Table 3. Possible applications of a-SiC.

1. Window for Solar Cells
2. Emitter for HBT
3. Passivation for Thermal Head etc.
4. Radiation Hardness and High Temperature Device
5. Sensors
6. Electrodes for Electrolyte

6. CONCLUSION

It is shown that a-SiC material is very useful for Si-related electronic devices. The reasons are 1) the fairly good electronic properties at the interface between it and crystalline Si, 2) it is a chemically stable material, 3) it is mechanically strong and 4) it is easily fabricated at relatively low temperature. At the moment, this material has problems such as 1) relatively high specific resistance 2) the degradation of the properties at high temperature (>600 °C). These two problems might be solved by introducing 1) micro-crystalline phase and 2) fluorine atoms instead of hydrogen atoms as the dangling bond terminaters.

ACKNOWLEDGMENTS

The author wishes to thank Assoc. Prof. H. Ishiwara, Dr. T. Asano, Messers K. Sasaki and T. Fukazawa for their useful discussions.

This work was partially supported by NHK Hoso-Bunka Foundation and CASIO Science Promotion Foundation.

REFERENCES

[1]K. Sasaki, M. M. Rahman and S. Furukawa : 1984 Fall Meeting of the Japan Society of Applied Physics, 14p-A-4, p.409 (1984).
[2]M. Ghannam, J. Nijs, R. Mertens and R. Dekeersmaecker : 1984 IEDM Tech. Dig., pp.746-748 (1984).
[3]K. Sasaki, M. M. Rahman and S. Furukawa : Technical Report of Inst. of Electronics and Communication Engineers of Japan, ED85-8, pp.15-20 (1985).
[4]M. M. Rahman and S. Furukawa : 1st Int. PVSEC, Tech. Dig., pp.171-174 (1984).
[5]K. Sasaki, M. M. Rahman : IEEE Elec. Dev. Lett., EDL-6[6], pp.311-312 (1985).
[6]K. Sasaki and S. Furukawa : Ext. Abs. 18th Int. Conf. SSDM, pp.291-294 (1985).
[7]K. Sasaki, T. Fukazawa, S. Furukawa : 1987 IEDM Tech. Dig., pp.186-189. (1987).

Fabrication of a-Si:H/a-SiC:H Multilayers by a Glow Discharge Method and Carrier Transport Properties

H. Matsunami, M. Yoshimoto, and T. Fuyuki

Department of Electrical Engineering, Kyoto University, Yoshidahonmachi, Sakyo, Kyoto 606, Japan

Multilayers of a-Si:H/a-$Si_{1-x}C_x$:H were fabricated using a glow discharge method from a mixture of SiH_4 and C_2H_4. The barrier layer using a-$Si_{1-x}C_x$:H was typically 20 Å thick, and the thickness of the well layer was varied from 11 to 510 Å. The configuration of the multilayer was confirmed by small angle X-ray diffraction. The hydrogen incorporation scheme in the well layer of a-Si:H was investigated by means of infrared absorption measurements. The structure of the a-Si:H layer was strongly influenced by its thickness. The electron transport properties across the multilayer were also studied. Tunneling current was observed when the barrier-layer thickness became narrow. Using a novel structure, it turned out that tunneling electrons were hot electrons.

1. Introduction

In recent years, there has been interest in multilayers consisting of hydrogenated amorphous silicon (a-Si:H) and silicon-based ternary alloys, e.g. a-SiN_x:H[1,2], a-$Si_{1-x}C_x$:H[3-5] and a-$Si_{1-x}Ge_x$:H [6,7] because of their unique optical[1-4] and electrical [1,2,5-7] properties. In multilayers with very thin layers, the optical energy gaps (E_{opt}) of the multilayers increase with the decreasing thickness of well layers, due to the quantum size effect [1-4]. There is also information in the literature on the electrical properties along multilayer planes[1,2]. In addition, a-Si:H/a-$Si_{1-x}C_x$:H multilayered systems have already been used effectively in solar cells as effective p-type window layers[6].

In this report, deposition of a-$Si_{1-x}C_x$:H is described as the base for fabrication of amorphous multilayers. In order to obtain multilayers with specific properties, it is important to control structures of component layers and interfaces. The microscopic structures of the a-Si:H/a-$Si_{1-x}C_x$:H (x=0.8) multilayer fabricated by a glow discharge method are discussed. The scheme of hydrogen incorporation in the well layer was determined by the analysis of infrared absorption measurements for the well-layer thickness over the range from 11 Å to 510 Å. We discuss the generation of hot electrons by tunneling conduction through barrier layers in a-Si:H/a-$Si_{0.2}C_{0.8}$:H multilayers. This fundamental result of hot electron conduction is important for the application of a-Si:H/a-$Si_{1-x}C_x$:H multilayers to a hot carrier injector or a wide gap injector.

Springer Proceedings in Physics, Vol. 34 Amorphous and Crystalline Silicon Carbide
Editors: G.L. Harris and C.Y.-W. Yang

2. Deposition of a-$Si_{1-x}C_x$:H

A-$Si_{1-x}C_x$:H films 0.5-1.0 microns thick were deposited on polished silicon wafers or glass substrates in a capacitive-coupling rf (13.56 MHz) glow discharge chamber[8] using a mixture of SiH_4, C_2H_4 and H_2. The deposition conditions were varied by changing the flow rate of the source gases. Input rf power was kept at 10 W. Gas pressure in the reaction chamber was maintained at 0.94-1.6 Torr. Substrate temperature (T_{sub}) ranged between 220 and 260°C, with room temperature (RT) as a reference. The carbon and silicon contents in the films were determined by Auger electron spectroscopic analysis, and the incorporated amount of hydrogen was measured using infrared spectroscopy (IR).

The average numbers of H atoms attached to C atoms, [C-H]/[C], in the films of T_{sub}=RT and 260 °C are shown in Fig.1. The number of C atoms in a unit volume of the sample, represented by [C] in Fig.1, was deduced by a method described in a previous paper [9,10]. The value [C-H], the numbers of H atoms bonding to C atoms, was estimated from IR measurements, using a value of 1.0×10^{21} cm^{-2} [11] as the inverse absorption cross section for C-H_n (n=1, 2 and 3) stretching modes. The values of [C-H]/[C] in the films are 0.5 for films of T_{sub}=260°C and 1.8 for T_{sub}=RT; the hydrogenation of C atoms is much larger than that of Si atoms[8]. In the carbon-rich films (x>0.8 for T_{sub}=RT and x>0.6 for T_{sub}=260°C), [C-H]/[C] steeply increases. Thus, the number of C atoms increases as C-H_2 or C-H_3 in carbon-rich films. The excess hydrogenation of C atoms increases flexibility of film structures. On the basis of the optical emission spectroscopy for the glow discharge plasma [9,12], reduction in the decomposition of C_2H_4 in the higher C_2H_4 gas composition causes the excess hydrogenation of C atoms.

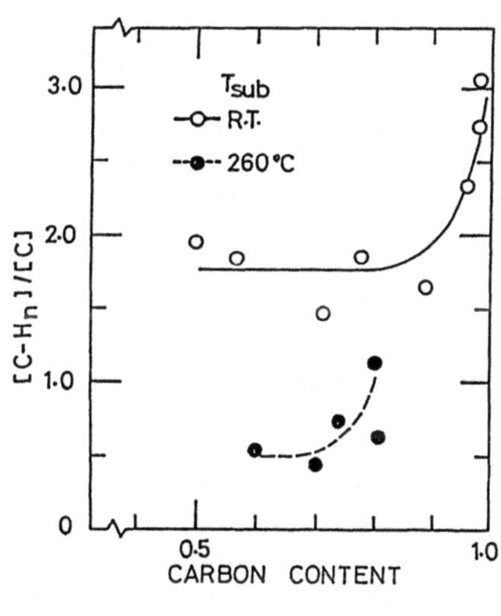

Fig. 1. Average number of H atoms attached to a C atom vs. carbon content.

The optical energy gap increases monotonically with the carbon content (Fig.2), and it does not show a maximum in contrast with the results obtained by other researchers[13,14]. Instead, E_{opt} increases more steeply for a carbon content above 0.8 for T_{sub}=RT and above 0.6 for T_{sub}=260 °C; this is due to a steep increase in the average number of hydrogen atoms attached to carbon atoms as shown in Fig.1. The maximum value of E_{opt} obtained is as high as 4.12 eV in a sample (T_{sub}=RT) with a carbon content of 0.97. At the same carbon content, E_{opt} for the film of T_{sub}=RT is larger than that of T_{sub}=260 °C; this is due to a larger hydrogen incorporation in the film of T_{sub}=RT.

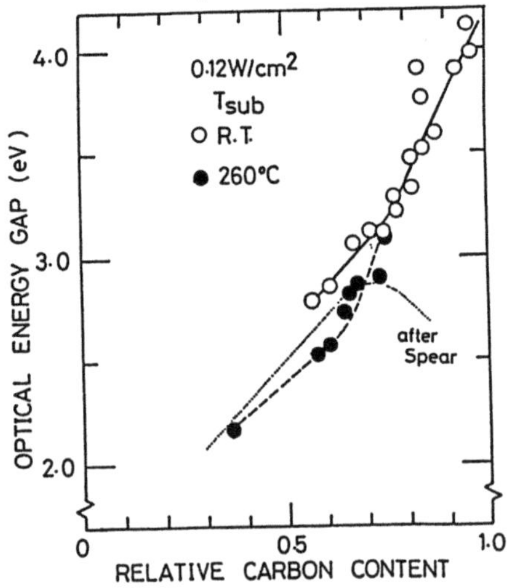

Fig. 2. Optical energy gap vs. carbon content. Also shown is the result from ref.13.

3. Fabrication of Multilayers

The multilayers (well: a-Si:H, barrier: a-$Si_{1-x}C_x$:H, x=0.8) were prepared by the glow discharge deposition with T_{sub}=220 °C. a-Si:H layers for the well layer were deposited from SiH_4 (10 % H_2 diluted). The thickness of the well layer (d_W) in the multilayer was varied from 11 to 510 Å by changing the deposition time. For thick a-Si:H (5000 Å), E_{opt} was 1.78 eV. The thickness of the barrier layer (d_B) was kept at 20 Å. For thick a-$Si_{1-x}C_x$:H (around 5000 Å), the optical energy gap was 3.18 eV for x=0.8. Neither a-Si:H nor a-$Si_{1-x}C_x$:H were doped.

To avoid carbon contamination in a-Si:H layers, the discharge was stopped at the end of each layer deposition and the source gases were completely changed. The period of the multilayer was maintained at 20, when the well layer was thinner than 100 Å. Above 100 Å, the period was kept at 7.

The construction of the multilayer was confirmed by Auger depth profile[15] and X-ray diffraction. Figure 3 shows an example of an X-ray diffraction pattern in a case where the thickness of the well layer and the barrier layer were 50 and 20 Å, respectively. The multilayer showed sharp peaks in the X-ray diffraction pattern. The full width at half maximum of the first Bragg peak in the 2θ scan was 0.08°. This result indicates that stratified layers were fabricated without fluctuation in thickness of the component layers. The thickness d_W+d_B was calculated from the X-ray diffraction peak. The deposition time for the barrier layer was kept constant. The value of d_W+d_B linearly varied with the deposition time for the well layer, which indicates that the fabrication of multilayers was controllable. The intercept with the ordinate indicates the thickness d_B (20.3 Å) and the slope shows the deposition rate of a-Si:H (1.69 Å/s).

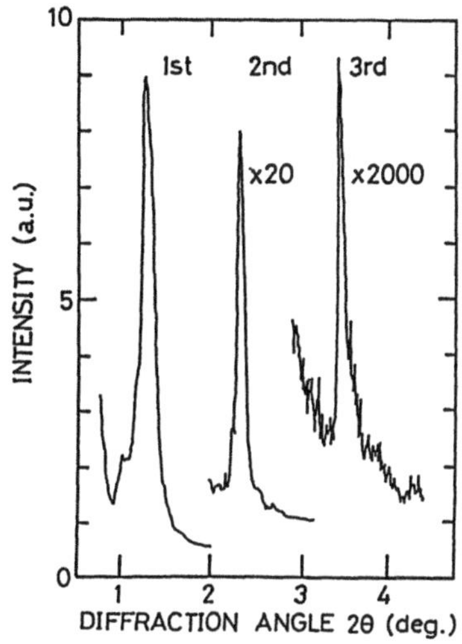

Fig. 3. X-ray diffraction pattern with 1st, 2nd and 3rd peaks in the case that the thickness of the well layer and the barrier layer were 50 and 20 Å, respectively. The carbon content of the barrier layer was 0.8.

4. Scheme of Hydrogen Incorporation

Hydrogen incorporation in the multilayer was investigated by Fourier transform infrared absorption spectroscopy (FTIR). In Fig. 4 are shown examples of absorption spectra for $Si\text{-}H_n$ stretching modes (n=1 and 2) in a-Si:H/a-$Si_{0.2}C_{0.8}$:H multilayers. In the same figure, the spectra of thick a-$Si_{0.2}C_{0.8}$:H (4900 Å) and a-Si:H (5000 Å) are shown as references. The peak at 2000 cm^{-1} is due to

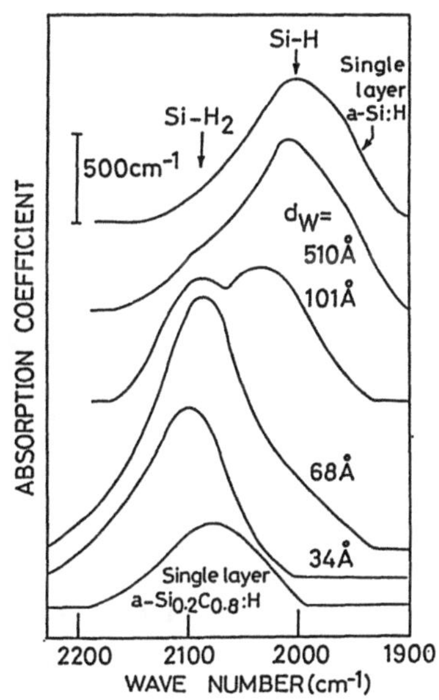

Fig. 4. Infrared absorption patterns of $Si\text{-}H_n$ (n=1 and 2) as a function of wave number for a-Si:H/a-$Si_{0.2}C_{0.8}$:H multilayers with various well-layer thickness; together with patterns for thick a-Si:H and a-$Si_{0.2}C_{0.8}$:H.

Si-H monohydride bonds in the a-Si:H layer. The peak around 2100 cm^{-1} is a mixture of Si-H monohydride bonds by C atoms in the a-$Si_{1-x}C_x$:H layer and $Si\text{-}H_2$ dihydride bonds in the a-Si:H layer.

Figure 5 shows the relations between the thickness of the well layer and the infrared absorption coefficient of the $Si-H_n$ stretching mode for the a-Si:H/a-$Si_{0.2}C_{0.8}$:H multilayer. In order to analyze the absorption quantitatively, the effective medium theory was used. A measured absorption coefficient α_M is expressed as follows[16-18]:

$$\alpha_M=\alpha_W\ (n_W/n_M)\ d_W/(d_W+d_B)+\alpha_B\ (n_B/n_M)\ d_B/(d_W+d_B), \quad (1)$$

where α_W and α_B are absorption coefficients for the well layer and the barrier layer, respectively. The peak absorption coefficients of the 2000 cm^{-1} mode for the well layer, denoted by closed circles in Fig.5, were calculated using Eq. (1). The refractive indices n_W and n_B were estimated from the measured refractive index in non-absorbing wavelength region.

This solid line for closed circles can be extrapolated to $d_W/(d_W+d_B)=0.66$ (d_W=38 Å). This result is explained by the fact that hydrogen atoms scarcely incorporate into the well layer in the form of Si-H bonds, when the well-layer thickness is thinner than 38 Å. Above 38 Å for the well-layer thickness, the number of Si-H bonds in a unit volume increases with the thickness of the well layer. This result is consistent with a two-dimensional nucleation process of a-Si:H observed by in situ ellipsometry measurements; islands merge on the scale of 35 Å and the properties of the growing film approach the bulk properties of a-Si:H[19].

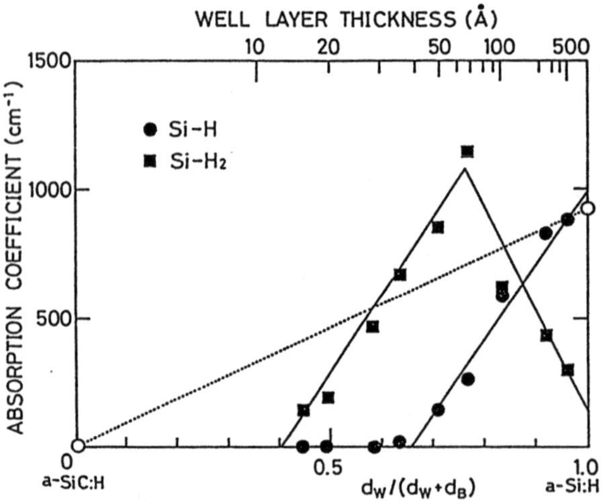

Fig. 5. Peak absorption coefficients of the 2000 cm^{-1} mode (closed circles) and the 2100 cm^{-1} mode (squares) for a-Si:H/a-$Si_{0.2}C_{0.8}$:H as a function of well-layer thickness, together with peak absorption coefficients of the 2000 cm^{-1} mode in a-Si:H and a-$Si_{0.2}C_{0.8}$:H (open circles).

In Fig.5 are also shown absorption coefficients of 2090 cm^{-1} mode for the well layer. The absorption coefficients were corrected by subtracting contribution of the monohydride bonds in the barrier layer (a-$Si_{0.2}C_{0.8}$:H), where α_W was estimated with Eq.(1) using n_B=1.7. The corrected values denoted by squares in the figure were deduced on the assumption that the number of Si-H monohydride bonds in a unit volume of the barrier layer is the same as in thick a-$Si_{0.2}C_{0.8}$:H (i.e., α_B in the barrier layer is the same as in thick a-$Si_{0.2}C_{0.8}$:H).

In the range of $d_W/(d_W+d_B)$ from 0.45 (d_W=16 Å) to 0.77 (d_W=65 Å), the solid line for squares linearly increases with $d_W/(d_W+d_B)$. The extrapolated value of the line to the abscissa is $d_W/(d_W+d_B)$= 0.39 (d_B=13 Å). Above 0.77 for $d_W/(d_W+d_B)$, the absorption coefficient of the SiH_2 mode decreases with increasing $d_W/(d_W+d_B)$. This result indicates that hydrogen atoms scarcely incorporated into the well layer in the form of Si-H_2 bonds, when it is thinner than 13 Å. In the range of the well-layer thickness from 13 Å to 65 Å, the number of Si-H_2 bonds in a unit volume of the well layer increases with the thickness. Above 65 Å, the number of Si-H_2 bonds decreases with the thickness. When the well-layer thickness is thinner than 100 Å, absorption coefficients of the 2090 cm^{-1} mode are still larger than that of the 2000 cm^{-1} mode as shown in Fig.5. Therefore, Si-H_2 dihydride bonds are dominant rather than Si-H monohydride bonds below d_W of 100 Å.

5. Electrical Properties in Multilayers

5-1. Current through Multi-heterojunction

Figure 6 shows typical I-V characteristics of the a-Si:H/a-$Si_{0.2}C_{0.8}$:H multilayer structure with two barrier layers. The structure was prepared on a NiCr predeposited Corning 7059 glass substrate. The a-Si:H was n-type (phosphorus-doped) with the dark conductivity of 7.4×10^{-3} S/cm and the activation energy of the conductivity of 0.26 eV. The thickness of the a-Si:H layer was 50 Å. The barrier layer of a-$Si_{0.2}C_{0.8}$:H was 30 Å thick. The upper electrode of Al was evaporated. The contact between phosphorus-doped a-Si:H and the NiCr or Al electrode was verified as ohmic.

The dark conductivity was measured across the multilayer plane at room temperature and at 90 K. The I-V curve was measured by applying dc voltage. The same results described below were obtained by applying pulse voltage of 150 msec width and 1 Hz repetition to avoid Joule heating. At room temperature, the current flows in Ohm's law with a conductivity of 2.1×10^{-12} S/cm below 0.15 V as shown in Fig.6. Above 0.15 V, the I-V characteristic deviates from Ohm's law, and then the current increases drastically.

Using the difference in conductivities between a-Si:H and a-$Si_{0.2}C_{0.8}$:H as mentioned above, the voltage is mainly applied to the a-$Si_{0.2}C_{0.8}$:H barrier layers. The field in the a-$Si_{0.2}C_{0.8}$:H barrier layer can be estimated to be 1×10^{6} V/cm at an applied voltage of 0.6 V where the current increases drastically. The I-V characteristic at 90 K is the same as at room temperature except for the decrease in the amount of current. The decrease in current is due to a decrease in the conductivity of the a-Si:H layer at 90 K.

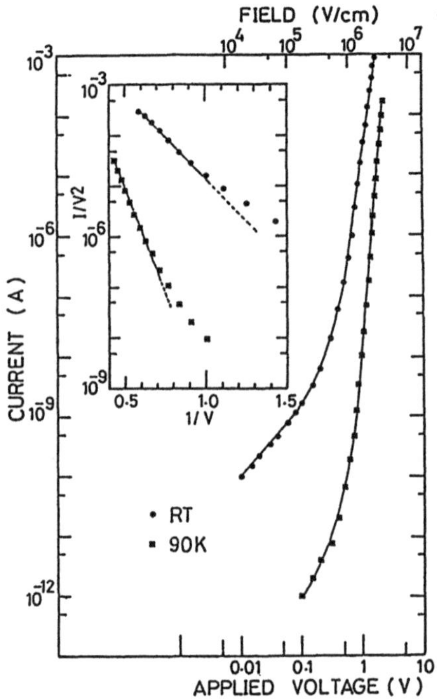

Fig. 6. I-V characteristics across the multilayer plotted logarithmically. The well layer of the multilayer is n-type. Inset shows I-V data plotted as $\log(I/V^2)$ vs. 1/V.

The inset of Fig.6 shows the I-V data plotted as $\log(I/V^2)$ vs. 1/V. The value of $\log(I/V^2)$ is in proportion to the inverse of the applied voltage above 1.0 V at both room temperature and at 90 K. The linear relationship of $\log(I/V^2)$ vs. 1/V is satisfied in the two-orders-of-magnitude range. It can be concluded that the current above 1.0 V is the tunneling current through a-$Si_{0.2}C_{0.8}$:H barrier layers, because the tunneling current is expressed as[20]

$$J \propto E^2\exp(-4\sqrt{2m^*}(2\phi_B)^{3/2}/3q\hbar E), \quad (2)$$

where J is the current density, E the applied field, q the charge of electron, $\hbar$ the Planck constant, m^* the effective mass of electron, ϕ_B the effective barrier height for tunneling. The value of 1.0×10^{6} V/cm above which the current increases drastically is comparable to the well-known value[21] for the onset of the tunneling current.

The effective barrier height ϕ_B can be estimated as 0.30 eV and 0.52 eV at RT and at 90 K, respectively, from the slope of the curve in the inset of Fig.6. The effective mass of an electron was assumed to equal the free-electron mass. The temperature dependence of the conductivity in a-Si:H indicated that hopping conduction is dominant at low temperature, i.e., 90 K. Thus, the energy level for the path of electrons shifted to the midgap from extended states, which resulted in the higher barrier height for tunneling conduction at low temperatures.

Tunneling conduction was not observed in the multilayer with the p-type well layer. However, space charge limited conduction was dominant.

5-2. Hot Electron Generation

Current-voltage characteristics were measured in a single heterojunction of phosphorus-doped a-Si:H (n-type) and $a\text{-}Si_{0.2}C_{0.8}:H$. In the heterojunction, the I-V characteristics showed ohmic conduction as shown in Fig.7. The conductivity was $2\sim4\times10^{-14}$ S/cm. This I-V characteristic is thought to be due to conduction of electrons through $a\text{-}Si_{1-x}C_x:H$, because n-type a-Si:H is thought as a source of carriers. The carrier injection from a-Si:H to the conduction band of $a\text{-}Si_{0.2}C_{0.8}:H$ does not occur in this single heterojunction, because there is a barrier for electrons in the conduction band[22].

The detection of hot electron generation was carried out using a newly devised structure shown in Fig.8. A double-barrier multilayer was fabricated on a thick $a\text{-}Si_{0.2}C_{0.8}:H$ bulk (1000 Å) with a thin buffer layer. The thickness d of the buffer layer between the multilayer and the bulk in the figure was changed. Current may flow as follows in the case where the $a\text{-}Si_{0.2}C_{0.8}:H$ bulk side is positively biased to the multilayer side, namely forward bias. (a) If transferred electrons by tunneling through the multilayer do not lose kinetic energy, i.e., electrons become "hot", the electrons can be injected to $a\text{-}Si_{0.2}C_{0.8}:H$ as shown in Fig.8(a). (b) If transferred electrons by tunneling lose kinetic energy by scattering in

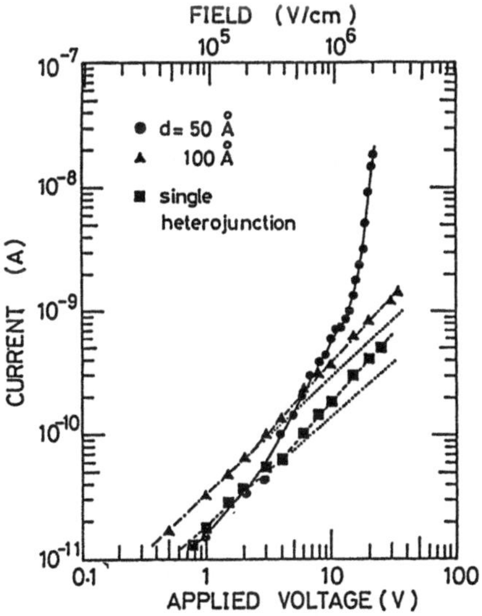

Fig. 7. I-V characteristics of devised structures and the single heterojunction.

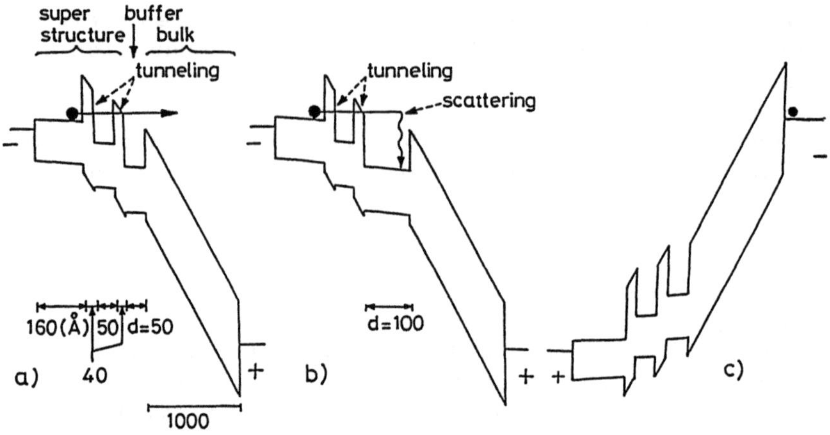

Fig. 8. Schematic electronic band diagram of devised structure; Forward biased:
(a) hot electron injection (d=50 Å),
(b) non-injection (d=100 Å).
Reverse biased: (c).

the buffer layer, the electrons cannot be injected to the conduction band of a-$Si_{0.2}C_{0.8}$:H as shown in Fig.8(b), because the electron injection does not occur in a a-Si:H/a-$Si_{0.2}C_{0.8}$:H single heterojunction as mentioned above. In the case where the a-$Si_{0.2}C_{0.8}$:H bulk side is negatively biased to the multilayer side, the I-V characteristics may be the same as in the single heterojunction because of the barrier for electron as shown in Fig.8(c).

In Fig.7, I-V curves of the devices with different d are shown with circles and triangles. In the case of d=50 Å, the current increased drastically above the applied voltage of 15 V giving an electric field of 1.4×10^{6}V/cm to the multilayer. The electric field corresponds to the onset of tunneling current through the multilayer. Above the applied voltage where the tunneling current starts to flow, hot electron injection occurs as shown in Fig.8(a). Drastic increase in the current was not observed when the multilayer side was positively biased to the a-$Si_{0.2}C_{0.8}$:H bulk side, i.e., reverse bias.

When the thickness d increased above 100 Å, electron injection was not observed as well in the case of the single heterojunction. The result implies that transferred electrons by tunneling through the multilayer are scattered in the buffer layer and lose kinetic energy for d=100 Å as shown in Fig.8(b).

6. Summary

(1) A-Si:H/a-$Si_{0.2}C_{0.8}$:H multilayers were fabricated by glow discharge method. The barrier layer using a-$Si_{0.2}C_{0.8}$:H was kept at 20

or 30 Å thick. The thickness of the well layer consisting of a-Si:H was changed from 11 Å to 510 Å. The construction of the multilayer was confirmed by X-ray diffraction and depth profile of Auger electron spectroscopy.

(2) Configuration of $Si-H_n$ (n=1 and 2) in the well layer (a-Si:H) was investigated by infrared absorption spectroscopy. Hydrogen atoms are incorporated as $Si-H_2$ dihydride bonds when the well-layer thickness is thinner than 38 Å. Above 38 Å for the well-layer thickness, Si-H monohydride bonds are formed. Up to 100 Å for the thickness of the well layer, $Si-H_2$ dihydride bonds are dominant rather than Si-H monohydride bonds.

(3) The tunneling current across the amorphous multilayer with the n-type well layer was demonstrated at room temperature. The barrier layer of a-$Si_{0.2}C_{0.8}$:H with the optical energy gap of 3.18 eV was typically 30 Å thick.

(4) The I-V characteristics were measured in the newly devised structure, and generation of hot electrons was verified in the a-Si:H/a-$Si_{0.2}C_{0.8}$:H multilayer with the n-type well layer. Above the applied voltage where tunneling current starts to flow through the multilayer, hot electron injection occurs through the buffer layer of 50 Å thick. This a-Si:H/a-$Si_{0.2}C_{0.8}$:H multilayer with a thin buffer layer can be used as a hot electron injector or a wide gap injector.

References

[1] B.Abels and T.Tiedje : Appl. Phys. Lett. **45** (1984) 179.

[2] N.Ibaraki and H.Fritzche : Phys. Rev. B **30** (1984) 5761.

[3] H.Munekata and H.Kukimoto : Jpn.J.Appl.Phys. **22** (1983) L544.

[4] T.Tiedje, B.Abels, P.D.Person, B.G.Brooks and G.D.Cody: J.Non-Cryst.Solids **66** (1984) 345.

[5] S.Tsuda, H.Tarui, T.Matsuyama, T.Takahama, S.Nakayama, Y.Hishikawa, N.Nakamura, T.Fukatsu, M.Ohnishi, S.Nakano and Y.Kuwano, Jpn.J.Appl.Phys. **26** (1987) 28.

[6] H.Shirai, A.Tanabe, S.Oda, J.Hanna, T.Nakamura and I.Shimizu: Appl.Phys. **A41** (1986) 259.

[7] C.R.Wronski, P.D.Persons and B.Abels: Appl.Phys.Lett. **49** (1986) 569.

[8] J.Saraie, Y.Fujii, M.Yoshimoto, K.Yamazoe and H.Matsunami: Thin Solid Films **117** (1984) 59.

[9] M.Yoshimoto, K.Aizawa, T.Fuyuki and H.Matsunami: Jpn.J.Appl. Phys. **25** (1986) 1465.

[10] M.Yoshimoto, T.Fuyuki and H.Matsunami: Jpn.J.Appl.Phys. **25** (1987) 996.

[11] F.Fujimoto, A.Ootuka, K.Komaki, Y.Iwata, I.Yamane, H.Yamashita, Y.Hashimoto, Y.Tawada, K.Nishimura, H.Okamoto and Y.Hamakawa: Jpn.J.Appl.Phys. **23** (1984) 810.

[12] H.Matsunami, M.Yoshimoto and T.Fuyuki: 7th Int. Symposium on Plasma Chem. Eindhoven, 1985 (IUPAC,1985)p.154.
[13] D.A.Anderson and W.E.Spear: Philos. Mag. **35** (1977) 1.
[14] R.Sussmann and R.Ogden: Philos. Mag. **44** (1981) 137.
[15] M.Yoshimoto, T.Fuyuki and H.Matsunami: Jpn.J.Appl.Phys. **25** (1986) L21.
[16] H.Ugur, R.Johanson and H.Fritzche: "Tetrahedrally-Bonded Amorphous Silicon Semiconductors" ed. D.Adler and H.Fritzche (Plenum, New York, 1985). p.425.
[17] B.Abels, L.Yang, P.D.Persons and H.S.Stasiewski and W.Lanford: Appl.Phys.Lett. **48** (1986) 168.
[18] M.Yoshimoto, T.Fuyuki and H.Matsunami: 8th Int. Symposium on Plasma Chem. Tokyo, 1987 (IUPAC,1987) p.1508.
[19] R.W.Collins and A.Pawlowski: J.Appl.Phys. **59** (1986) 1160.
[20] J.G.Simmons : J. Appl. Phys. **34** (1963) 1793.
[21] S.M.Sze : "Physics of Semiconductor Devices" (John Wiley & Sons, New York, 1981) 2nd ed., p.402
[22] M.Yoshimoto, T.Fuyuki and H.Matsunami: Jpn.J.Appl.Phys. **25** (1986) L992.

Highly Photosensitive a-SiC:H Films Prepared at High Deposition Rate by Glow Discharge in SiH_4-C_2H_2 Mixture Gas

S. Akita, K. Wakita, Y. Nakayama, and T. Kawamura

College of Engineering, University of Osaka Prefecture,
Sakai, Osaka 591, Japan

Highly photosensitive a-SiC:H films exhibiting $\eta\mu\tau$-product more than $10^{-6}cm^2/V$ at the band gap of 1.9 eV are prepared under a high deposition rate around 6μm/h by glow discharge using SiH_4-C_2H_2 source mixture gas. These films show high refractive indices, high $\sqrt{B}$-values and IR absorption spectra with a strong peak of Si-H centered near $2000cm^{-1}$ and almost no peak of C-H at $2850\sim2900cm^{-1}$. It is also shown that the SiH_4-C_2H_2 gas system has a feature of the carbon incorporation into the films with the content nearly equal to that in the source mixture gas.

1. INTRODUCTION

Hydrogenated amorphous silicon carbide (a-SiC:H) has been successfully used for a window layer of a-Si solar cells as a wide band-gap material and also for a surface protecting layer of a-Si photoreceptors as a material with high hardness and hydrophobia in addition to wide band-gap. As compared with hydrogenated amorphous silicon (a-Si:H) films, a-SiC:H films have poor photosensitivity at long wavelength (red) region because of the wide band-gap. From this opto-electric property, the application of a-SiC:H to the photoreceptors as a photoconductive layer with 20-30μm thickness is expected to design copy machines without a red cut filter which is usually used for a-Si photoreceptors. In order to realize this, a low deposition rate (<1μm/h) of highly photosensitive a-SiC:H films should be overcome. The purpose of this paper is to investigate highly photosensitive a-SiC:H films under a high deposition rate by glow discharge using SiH_4-C_2H_2 gas system.

2. EXPERIMENT

Amorphous SiC:H films were prepared by a capacitively coupled plasma deposition system. The reaction chamber had two cylindrical electrodes concentrically. A power supply (13.56MHz) was connected to the outer electrode. Substrates such as Al plates, Corning 7059 glasses, quartz glasses and crystalline Si wafers were attached to the inner electrode (108mm in diameter and 500mm length) grounded, held at 250°C and rotated at a rate of 10rpm.

A source mixture gas of SiH_4 and C_2H_2 was introduced together with H_2 into the chamber. The reason why C_2H_2 was chosen as a C source is that C_2H_2 is expected to be dissociated at a relatively low energy in the plasma as compared with other hydrocarbons such as C_2H_4, CH_4 and C_3H_8, because the thermal dissociation energy increases in the sequence C_2H_2, C_2H_4, CH_4 and C_3H_8. In fact, we have confirmed that the deposition rate of a-SiC:H films at almost the same deposition condition decreases in that order [1]. The optimization of the deposition parameters as rf power, pressure and H_2 flow rate was done under a relatively high SiH_4 flow rate.

The resultant films were characterized by measuring deposition rate, C content by x-ray microanalysis, ESR spectra, IR spectra, hydrogen content, refractive index, optical band-gap, sharpness of an optical absorption edge and dark and photoconductivities.

3. RESULTS AND DISCUSSION

Figure 1 shows the composition x in a-$Si_{1-x}C_x$:H films plotted against the C fraction $2C_2H_2/(SiH_4+2C_2H_2)$ in the source gas. The rf power, pressure, H_2 flow rate and total flow rate of SiH_4 and C_2H_2 are 600W, 1.2Torr, 300sccm and 315sccm, respectively. This is an optimized condition for $2C_2H_2/(SiH_4+2C_2H_2)=0.1$ in terms of high photosensitivity, which will be described later. In the figure the data of another series reported earlier [2] which were prepared under the rf power 150W, pressure 0.45Torr and SiH_4 flow rate 100sccm are also plotted. It is found that the C content x in the film almost equals the C fraction in the source gas. We believe that this is a feature of SiH_4-C_2H_2 gas system and is established for wide deposition conditions.

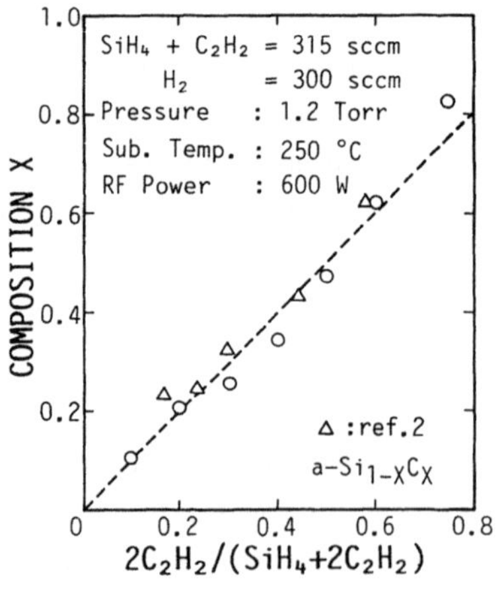

FIG.1. Composition x in a-$Si_{1-x}C_x$:H films plotted against the C fraction in the source gas.

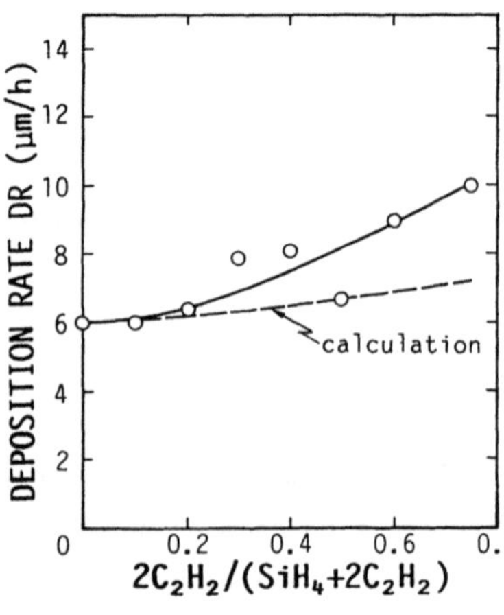

FIG.2. Deposition rate DR plotted against the C fraction in the source gas.

Figure 2 shows the deposition rate DR plotted against the C fraction in the source gas. The dotted line indicates a calculated curve considering the bond length of Si and C, and the value at zero C fraction is set to fit the experimentally obtained value of 6µm/h. On increasing the C fraction in the source gas beyond 0.2, DR increases away from the dotted line, which implies that the packing density of the film becomes low at high C fraction. It is noticed that DR as high as 6µm/h is obtained with high packing density up to the C fraction of 0.1.

Figure 3 shows the g-value in the ESR-spectra and the spin density N_s plotted against the composition x. The value of N_s increases with increasing x beyond 0.1 at which N_s is lowered by the optimization of deposition condition. The g-value does not change and has a value typical for Si dangling bonds (g=2.0055) up to a C content of x=0.3. Above this value of x the g-value gradually decreases and approaches a value for C dangling bonds (g=2.003), indicating that the increase in N_s at higher x comes from the defects due to both Si- and C-dangling bonds.

Figure 4 shows the contents of hydrogen bonded to Si $N_{H(Si)}$ and C $N_{H(C)}$, which are determined from Si-H (near 2000cm^{-1}) and C-H (near 2900cm^{-1})[3] stretching modes in the IR-spectra, respectively, and the wavenumber ν_{Si-H} of a peak near 2000cm^{-1} plotted against the composition x. With increasing x up to 0.5 the values of not only $N_{H(C)}$ but also $N_{H(Si)}$ increase. At x>0.6 there is a decrease in $N_{H(Si)}$ and in addition $N_{H(C)}$ becomes larger than $N_{H(Si)}$. The phenomenon of the increase in Si-H bonds with x, which has been reported even for the SiH_4-CH_4 gas system [4,5], has not been observed for the different deposition conditions [2]. The

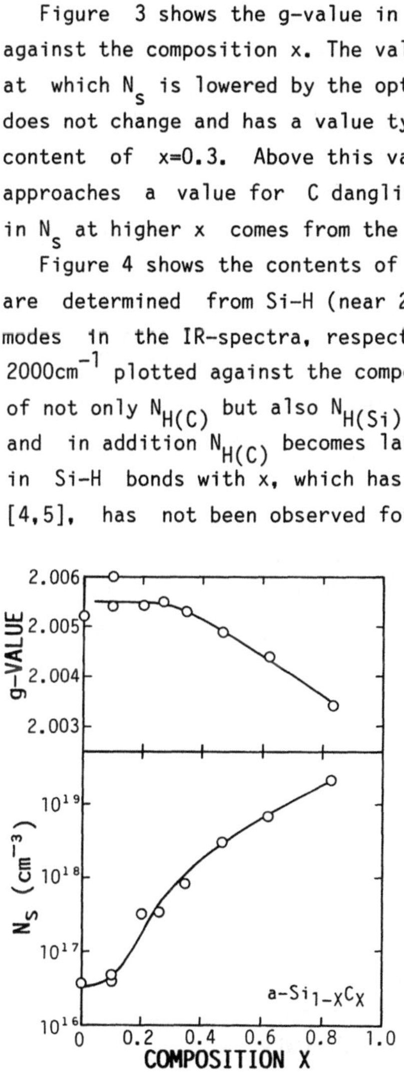

FIG.3. Film composition dependence of the g-value and the spin density N_s.

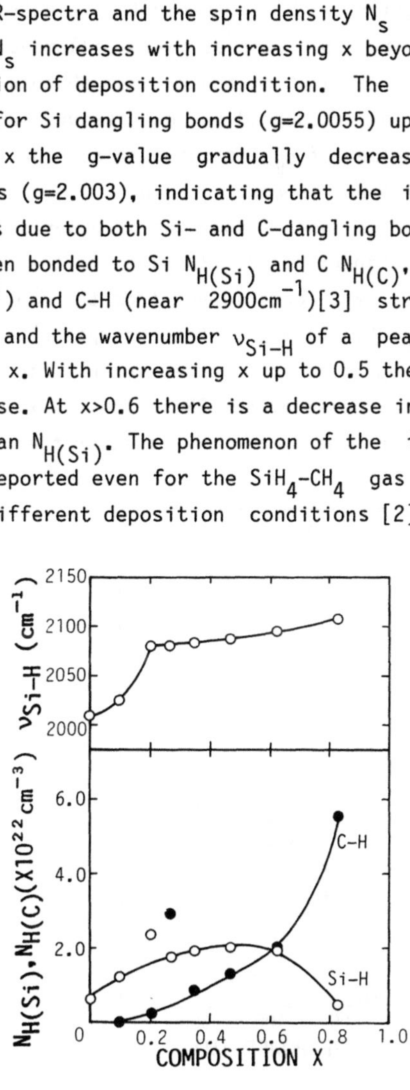

FIG.4. Film composition dependence of the wavenumber ν_{Si-H} and the contents of hydrogen bonded to Si $N_{H(Si)}$ and C $N_{H(C)}$.

wavenumber ν_{Si-H} is close to $2000cm^{-1}$ at $x \leq 0.1$ and to $2100cm^{-1}$ at $x \geq 0.2$. The shift of ν_{Si-H} from 2025 to $2087cm^{-1}$ at the range of x from 0.1 to 0.5 seems to come from the formation of SiH_2 bonds in addition to the back bonding of C to SiH.

For the fixed condition of $2C_2H_2/(SiH_4+2C_2H_2)=0.1$ other parameters such as rf power, pressure and H_2 flow rate are optimized in terms of photosensitivity. Figure 5 shows the dark and photoconductivities, σ_d and σ_{ph}, plotted against the rf power, where the rf power range from 50 to 600W corresponds to the power density from 0.02 to $0.24W/cm^2$. For the conductivity measurement a sample configuration with inter-digital electrodes was used, and for the photoconductivity the samples were illuminated by the light of $50\mu W/cm^2$ at the wavelength exhibiting the maximum photocurrent. As shown in the figure, both σ_d and σ_{ph} do not depend on the pressure but strongly on the rf power. Though the higher value of σ_{ph} is obtained at high rf power region, the maximum value of usable rf power is restricted by the peeling of the film.

Figures 6 and 7 show the rf power dependence of $N_{H(Si)}$, ν_{Si-H}, X, DR, the refractive index n, the $\sqrt{B}$-value and the optical band gap E_{gopt}. The values of $\sqrt{B}$ and E_{gopt} are determined from the $\sqrt{\alpha h\nu}-h\nu$ plot using the relation of $\alpha h\nu = B(h\nu - E_{gopt})^2$, where α is the absorption coefficient and $h\nu$ the photon energy. The value of $N_{H(C)}$ is not plotted in this figure because it is so small ($<5x10^{20}cm^{-3}$). At low rf power of 50W DR is low and x is higher than 0.1, indicating that the dissociation energy of C_2H_2 is lower than that of SiH_4. Above 150W, DR and x do not

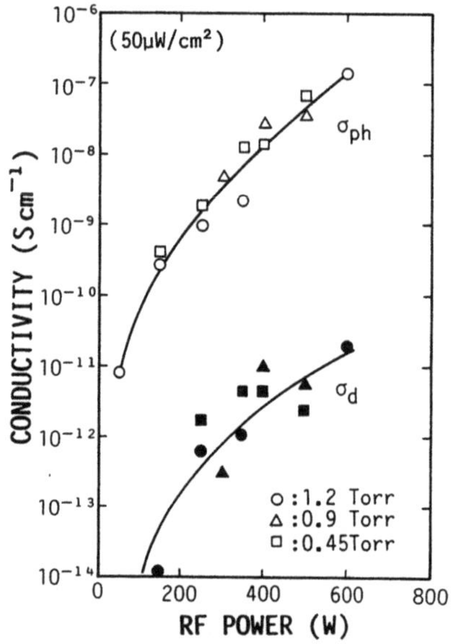

FIG.5. Rf power dependence of the dark and photoconductivities σ_d and σ_{ph}.

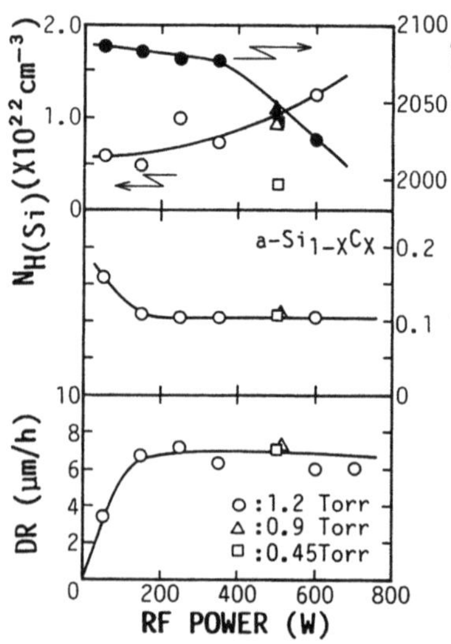

FIG.6. Rf power dependence of the hydrogen content $N_{H(Si)}$, the wavenumber ν_{Si-H}, the composition x and the deposition rate DR.

change. However, this is not due to the limitation of the source gas, because the emission intensities from SiH^* and CH^* do not saturate for the rf power range. It might be originated from the existence of considerable gas phase reaction between molecules and/or radicals including C atoms and those including Si atoms, leading to the enhancement of DR at lower rf power region reported previously [2]. At higher rf power region, though DR does not increase, $N_{H(Si)}$ and n (related to a film density) increase and ν_{Si-H} shifts toward $2000cm^{-1}$ with the rf power. These variations are desirable for achieving the high photosensitivity. The $\sqrt{B}$-value indicating the sharpness of the optical absorption edge also shows high value at 500∿600W. While, E_{gopt} decreases with the rf power, though the hydrogen content increases under no change of x. It is also found that the film properties described above do not depend on the H_2 flow rate ranging from 150 to 700sccm. At the H_2 flow rate lower than 100sccm, we could not obtain the films with a mirror-like surface.

Figure 8 shows σ_d and σ_{ph} plotted against E_{gopt} for a-Si:H films and a-SiC:H films, where samples denoted by a circle and a triangle are deposited at respective rates of 4∿6 and ∿1μm/h under various conditions. The a-SiC:H films with excellent photoconductivity of $\sigma_{ph}>10^{-7}Scm^{-1}$ ($\eta\mu\tau$-product$>10^{-6}cm^2/V$) at E_{gopt} of 1.9eV are obtained, as described above, at the high DR around 6μm/h available for fabricating thick-film devices. These films show the maximum of σ_{ph} at the wavelength of 550nm, which enables one to control the spectral response of photoreceptors as expected. In the case of low DR of ∿1μm/h we have obtained a-SiC:H films with

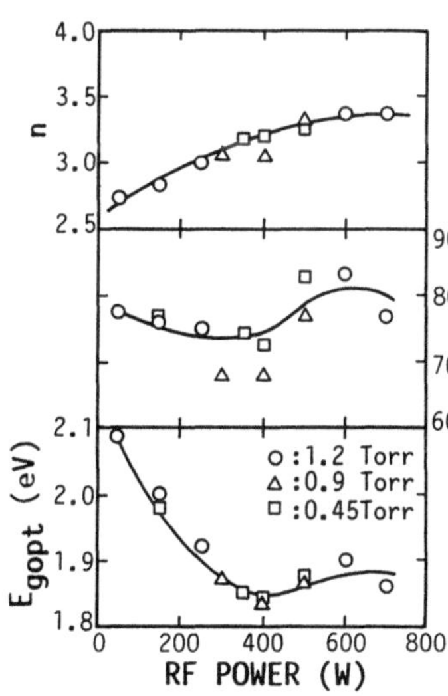

FIG.7. Rf power dependence of the refractive index n, the $\sqrt{B}$-value and the optical band-gap E_{gopt}.

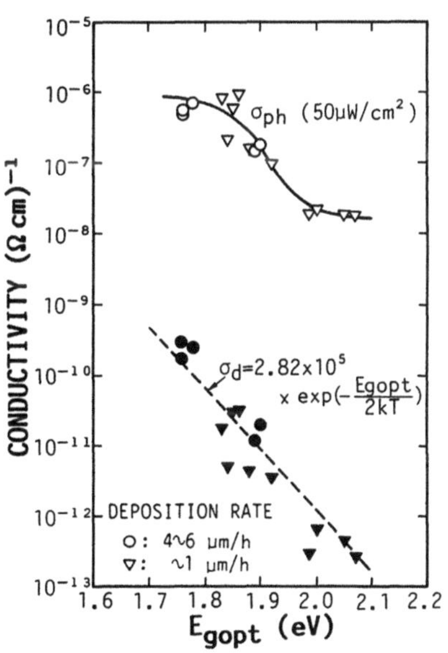

FIG.8. Dark and photoconductivities σ_d and σ_{ph} plotted against the optical band-gap E_{gopt}.

$\sigma_{ph}>10^{-8}Scm^{-1}$ ($\eta\mu\tau>10^{-7}cm^2/V$) up to E_{gopt}= 2.07eV. The data of σ_d for all samples almost fall on a line given by

$$\sigma_d = \sigma_0 \exp(-E_{gopt}/2kT),$$

where $\sigma_0=2.82\times10^5 Scm^{-1}$ and T=290K are used. The fitting indicates that σ_d varies corresponding to the value of E_{gopt}.

In conclusion, we have investigated the highly photosensitive a-SiC:H films with $\eta\mu\tau$-product$>10^{-6}cm^2/V$ at E_{gopt} of 1.9eV under the high deposition rate around 6µm/h. The films are prominent materials for photosensitive devices, especially for photoreceptors.

ACKNOWLEDGEMENTS

The authors thank Dr. S. Nakano and F. Ishikawa for the ESR measurements and N. Miyamoto for the carbon content measurements.

REFERENCES

1. M. Kudoh, M. Nakano, Y. Nakayama and T. Kawamura, Proc. SID, 28 (1987) 75.
2. Y. Nakayama, S. Akita, M. Nakano and T. Kawamura, Proc. 12th Int. Conf. on Amorphous & Liquid Semiconductors (Praha, 1987).
3. Y. Tawada, K. Tsuge, M. Kondo, H. Okamoto and Y. Hamakawa, J. Appl. Phys. 53 (1982) 5273.
4. A. Morimoto, T. Miura, M. Kumeda and T. Shimizu, Jpn. J. Appl. Phys. 21 (1982) L119.
5. G. H. Bauer, H. D. Mohring, G. Bilger and A. Eicke, J. Non-Cryst. Solids 77&78 (1985) 873.

A Novel *L*-Coupled RF PECVD System for Large-Area Deposition of a-SiC:H for Device Applications

A. Tran, D. Fung, M.M. Rahman, and C.Y. Yang

Microelectronics Laboratory, Santa Clara University, CA 95053, USA

A simple, L-coupled PECVD system capable of 4-inch fabrication of a-SiC:H films has been presented. We have designed and constructed a system that incorporates special features which are expected to enhance film quality and electronic and optical properties. Features such as precise temperature control of the substrate and flexible RF coil position are built into the system to optimize film uniformity and integrity. High-quality amorphous SiC:H films are deposited on various substrates for electrical and materials characterization measurements. Potentials of the resulting structures for device applications are assessed.

1. INTRODUCTION

Although hydrogenated amorphous silicon carbon alloy (a-$Si_{1-x}C_x$:H) films do not possess comparable electronic properties as crystalline SiC (which involves a high-temperature process), they have shown promising prospects in several device applications, such as solar cells [1,2], HBTs [3], etc. The two most significant advantages offered by a-SiC:H are low-temperature processing and controllable optical band-gap. However, the optical and electrical properties of a-SiC:H films depend significantly on the deposition method. Methods employed to produce a-SiC:H films include: RF glow discharge, reactive sputtering, ion cluster beam deposition, ion beam sputtering, and microwave discharge, etc. Figure 1 illustrates a few of these techniques. In particular, RF glow discharge Plasma Enhanced Chemical Vapor Deposition (PECVD) technique has been successful in producing high-quality a-SiC:H films with a high degree of reliability and integrity. There exists a large variety of PECVD system configurations. These include: planar, L-coupled tube, C-coupled tube, production hot-tube, and production radial-flow, etc. [4]. In the L-coupled system, the plasma is excited by a magnetic field; whereas in the C-coupled system it is achieved by an electric field. The conditions of glow discharge generated by these two systems are different, thereby resulting in different film characteristics. Moreover, the two types of PECVD reactors used in production environments are C-coupled and they utilize only one of the many plasma phenomena. Therefore, it seems reasonable to expect that future production reactors will take advantage of other phenomena to enhance film quality [4].

It has been reported that the application of a magnetic field to the glow discharge plasma confines the plasma, resulting in a reduction of the wall-plasma interaction, thus reducing contamination from the reactor wall [5]. Moreover, high doping efficiencies of boron and phosphorus atoms in a-Si:H have been achieved in such systems. It is reasonable to expect similar effects in an L-coupled PECVD system since it employs a magnetic field. Further, an L-coupled system can incorporate a provision for

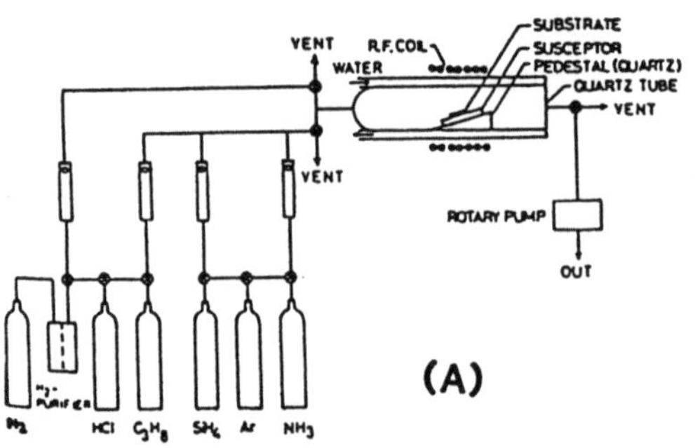

DEPOSITION RATE = 3-6 μm/HR

SUBSTRATE TEMP. = 1360 °C

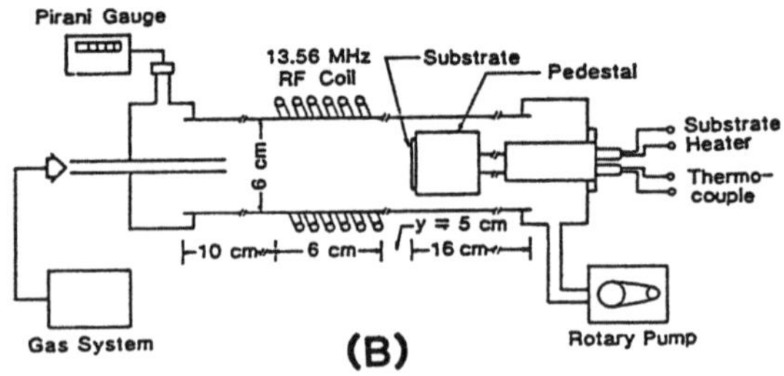

DEPOSITION RATE = 20 Å/MIN

SUBSTRATE TEMP. = 350 °C

DARK CONDUCTIVITY = 10^{-2} S/cm

R.F. POWER = 60 W

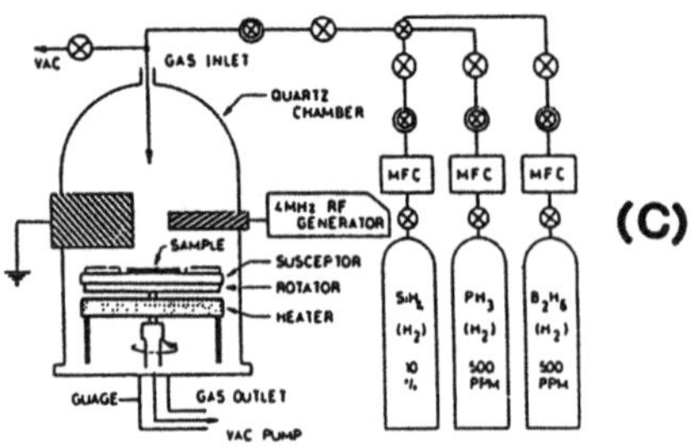

DEPOSITION RATE = 1 Å/SEC

SUBSTRATE TEMP. = 25-400 °C

DARK CONDUCTIVITY = 10^{-4} S/cm

R.F. POWER = 100 W

Fig.1 Comparison of several CVD systems

changing the relative positions of the substrate, the RF coil and the gas inlet. This would result in improved control of composition and concentration of reactive species in the plasma without altering macroscopic discharge parameters [6]. As a direct consequence, higher conductivity can be obtained. An L-coupled system is expected to give uniform films by virtue of the configuration of the system. In a previous report, amorphous SiC:H films, prepared by L-coupled PECVD technique, have been found to be of excellent glassy nature and adhered well to the substrate (quartz or Si) with good uniformity [7].

The above facts have encouraged us to design and construct an L-coupled PECVD system to fabricate a-SiC:H films in order to study their optical, electronic, and materials properties to ascertain their potential in device application. This paper presents some salient features of this system.

2. SYSTEM REQUIREMENTS AND DESIGN

Figure 2 shows a schematic representation of the system. Silane, methane, and dopant gases are continuously fed into a mixing tank at a predetermined ratio monitored by mass flow controllers in each line. The mixing tank provides a means for uniform mixing of the component gases. This is important given the fact that the resulting film is a compound semiconductor. The mixed gas is then introduced into the deposition chamber. Figure 3 shows a schematic

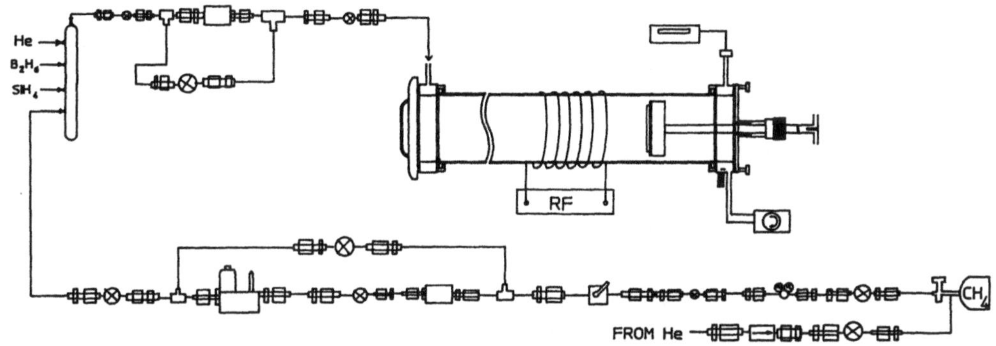

Fig.2 Overall schematics of the L-coupled PECVD system

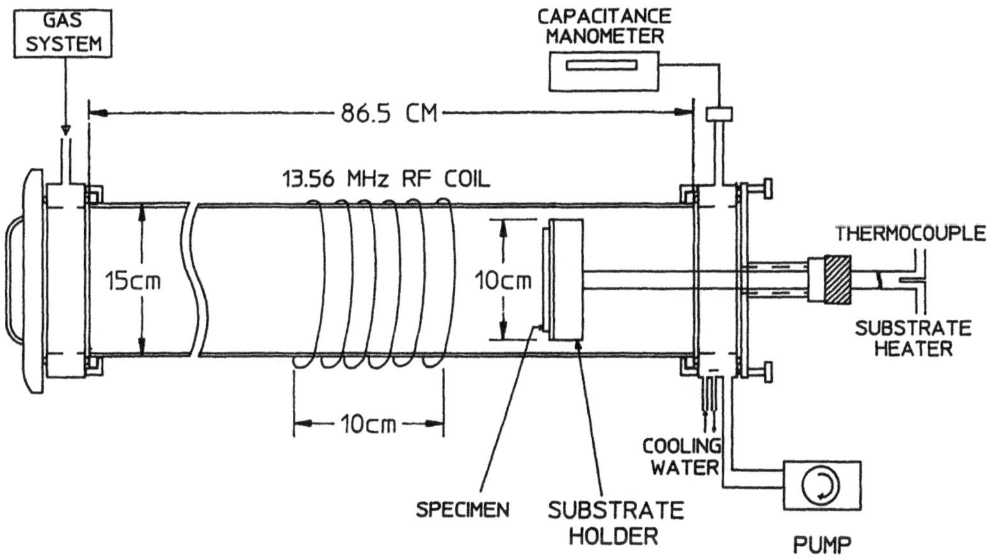

Fig.3 Reaction chamber schematic of the L-coupled PECVD system

diagram of the deposition chamber. The chamber has been designed to incorporate the following special features:

- It allows deposition on substrates up to 4 inches in diameter.
- Uniform temperature gradient on the substrate surface can be achieved.
- The relative positions of the gas inlet, the RF coil, and the substrate are made variable.
- Uniform cooling of substrate after deposition.
- Provision for a removable 5 inch inner tube inserted inside a fixed 6 inch reactor tube during deposition. This will minimize reactor-wall contamination.
- Expected laminar flow of gas mixtures in the chamber due to its large dimensions.

3. REPRESENTATIVE FILM PROPERTIES

Films deposited in a similar but smaller scale (2-inch) L-coupled system have a typical deposition rate of 0.5 Å/s. Some properties

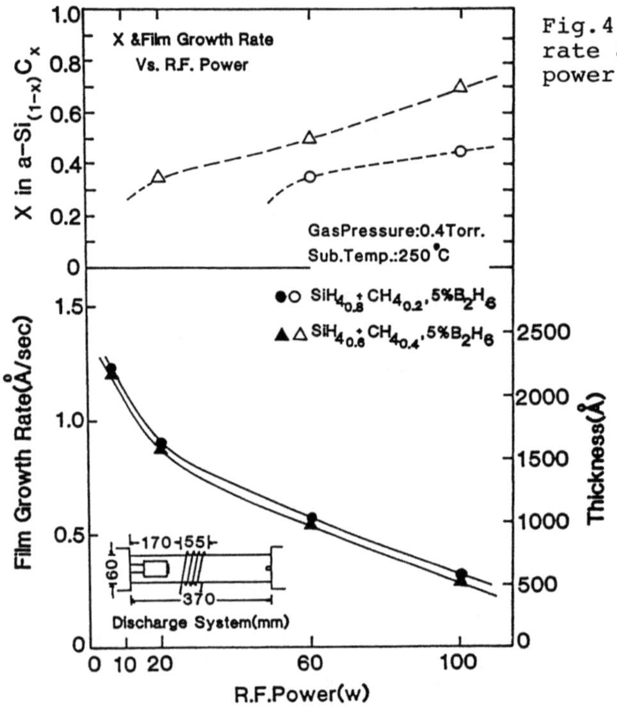

Fig.4 Dependence of film growth rate and x in a-$Si_{1-x}C_x$ on RF power

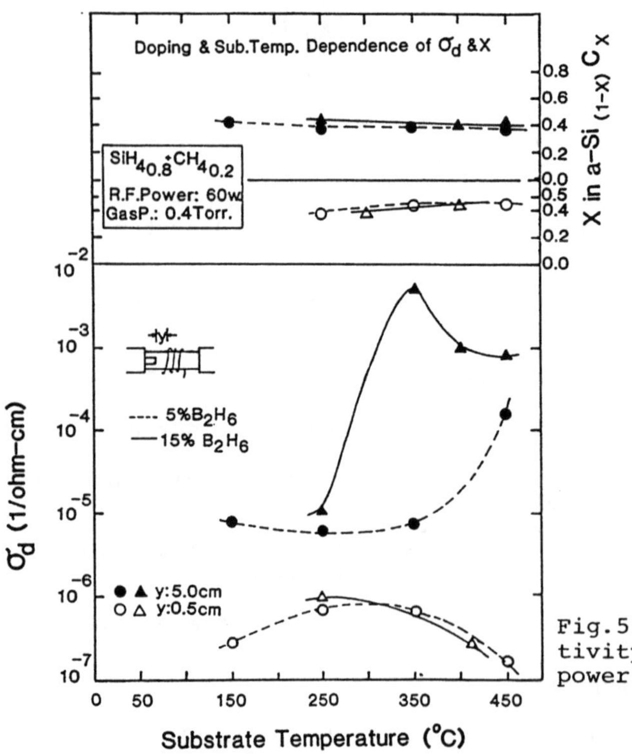

Fig.5 Dependence of dark conductivity and x in a-$Si_{1-x}C_x$ on RF power

of these films appear in Figures 4 and 5. An increase in the RF power caused carbon content in the film to increase, decreasing the film dark conductivity due to increasing bandgap. A dark conductivity of 10^{-2} S/cm for the films having x = 0.4 has been achieved by varying the relative positions of the RF coil, the substrate and the gas inlet, and increasing substrate temperature and gas fraction as shown in Fig. 5 for boron doped films [7]. Better film properties are expected in the 4 inch system.

4. SUMMARY

A simple, L-coupled PECVD system capable of 4-inch fabrication of a-SiC:H films has been presented. The system incorporates special features which are expected to enhance film quality and electronic and optical properties. An attempt is being made to study the impact of these design features on film properties and device application potentials.

REFERENCES

1. Y. Tawada, M. Kondo, H. Okamoto, and Y. Hamakawa: 'a-SiC:H/a-Si:H Heterojunction Solar Cell Having More Than 7.5% Conversion Efficiency,' Proc. 15th IEEE Photovoltaic Specialist Conf., Florida, (1981) p. 245.

2. M.M Rahman and S. Furukawa: 'Preparation and Performance of a Novel Structure Amorphous SiC/Crystalline Si Solar Cell,' Proc. 1st Int. Conf. PVSEC, Kobe (1984) p. 171.

3. K. Sasaki, S. Furukawa, and M.M. Rahman: Tech. Dig. 85 IEDM (1985) p. 294.

4. A. Sherman: 'Plasma-assisted Chemical Vapor Deposition Processes and their Semiconductor Applications,' Thin Solid Films, Vol. 113 (1984) p. 135.

5. M. Hirose: 'Some New Fabrication Technologies of Amorphous Si,' in Amorphous Semiconductor Technologies, ed. Y. Hamakawa, Ohm-sha, Tokyo (1982) p. 34.

6. K. Yokota, S. Nakayama, Y. Ohno, and S. Katayama: 'Growth of Hydrogenated Amorphous Silicon by and Inductive Coupling Glow Discharge Reactor,' Jpn. J. Appl. Physics, Vol. 26 (1987) p. 965.

7. M.M. Rahman and S. Furukawa: 'Electrical and Optical Properties of Amorphous SiC Films by Plasma CVD,' Proc. 8th Symp. Ion Sources and Ion-Assisted Technology, Tokyo (1984) p. 333.

Part III

Characterization of Silicon Carbide

Donor Identification in Thin Film Cubic SiC

S.G. Bishop[1], *J.A. Freitas, Jr.*[2], *T.A. Kennedy*[1], *W.E. Carlos*[1], *W.J. Moore*[1], *P.E.R. Nordquist, Jr.*[1], *and M.L. Gipe*[1]

[1] Naval Research Laboratory, Washington, DC 20375, USA
[2] Sachs/Freeman Associates, Inc., Landover, MD 20785, USA

Spectroscopic evidence for the presence of N in cubic SiC is reviewed, including earlier studies of Lely-grown crystals and recent investigations of thin films grown on Si substrates by chemical vapor deposition. This evidence and the results of new electron paramagnetic resonance, photoluminescence, and optically detected magnetic resonance studies of n-type films grown for various C/Si source gas ratios and p-type Al-doped films support the conclusion that there exists a residual donor in cubic SiC with binding energy of 54 meV which is attributable to isolated N impurities substituted on C sites. If the 15-20 meV donor observed in transport measurements is related to N, it must be in association with other defects (i.e. complexes) or in inhomogeneous regions of the film where the N is present at concentrations $\sim 10^{18} cm^{-3}$ or higher and can no longer be considered as isolated.

1. Introduction

Undoped films of cubic SiC grown epitaxially on Si substrates by chemical vapor deposition (CVD) are invariably strongly n-type, with the dominant residual donor having a binding energy in the range 15-20 meV as determined by Hall effect [1-4]. Photoluminescence (PL) spectra of such films [5,6] usually exhibit sharp line spectra attributed to excitons bound to neutral nitrogen donors [7], and electron paramagnetic resonance (EPR) studies [4] have detected neutral nitrogen donors, leading to the hypothesis that nitrogen is the dominant residual donor. However, the binding energy for the donor giving rise to the PL, as inferred from the PL spectra by two independent techniques [4,8-10], is about 54 meV, raising obvious questions concerning the identity of the 15-20 meV donor. Because the Hall measurements indicate significant compensation [1-4] there are additional questions concerning the existence of a compensating background acceptor and its identity.

In order to address these questions, it is useful to review the EPR evidence for the presence of nitrogen in SiC and the evidence for the assignment of bound exciton PL spectra to N donors [7,8] and donor-acceptor pair (DAP) PL spectra to nitrogen-aluminum pairs [9,10] in Lely-grown cubic SiC. Following this brief review, the current understanding of the role of N in CVD films of cubic SiC will be discussed from the perspective of recent PL, EPR, and optically detected magnetic resonance (ODMR) studies of the thin film material.

2. EPR and Photoluminescence Spectroscopy of Lely-Grown Cubic SiC

A neutral single donor, that is, a donor which has bound an electron, has an unpaired electron spin and can exhibit an EPR signal. The EPR spectrum can provide an identification of the donor

Springer Proceedings in Physics, Vol. 34 **Amorphous and Crystalline Silicon Carbide**
Editors: G.L. Harris and C.Y.-W. Yang

through the interaction of the paramagnetic electron with the nuclear magnetic moment of the donor atom [11]. This central hyperfine interaction splits the EPR band into 2I+1 lines, where I is the nuclear spin which can identify the donor. The strength of the hyperfine interaction provides a measure of the extent of the localization of the electron wavefunction at the nuclear site, and its dependence upon the relative orientation of the applied magnetic field and the crystal axes can reveal the site symmetry. In some cases, an interaction with the nuclear magnetic moments of the ligands (the superhyperfine interaction), can identify the neighboring atoms and thereby the sublattice on which the donor is located. In addition, a calibrated measurement of the EPR intensity determines the concentration of neutral donors which are present.

The earliest studies of electron paramagnetic resonance (EPR) in SiC were carried out by Woodbury and Ludwig [12] on N- and B-doped crystals of the 6H polytype. In the B-doped material they observed a ^{11}B- ^{29}Si ligand hyperfine interaction from which it was concluded that B impurities are located on C sites. In the N-doped crystals, for low N concentrations ($\sim 10^{16} cm^{-3}$), they observed a characteristic 3-line spectrum attributed to the central hyperfine interaction of the paramagnetic electron with the magnetic moment of the ^{14}N nucleus (nuclear spin I = 1). In this case no ligand hyperfine interaction was observed so it was not possible to determine on which sublattice the N is located. However, the 3-line EPR spectrum was established as a fingerprint for the presence of neutral N donors.

Following additional studies of the temperature dependence and the exchange interaction for EPR in N-doped 6H SiC by Veinger [13] and Zaritskii et al. [14], Altaiskii et al. [15] reported the first EPR spectra for N-doped cubic SiC. The 3-line EPR spectrum characteristic of the ^{14}N central hyperfine interaction was observed, and on the basis of the magnitude of the hyperfine splitting it was determined that the N donor in the cubic material has a larger orbit than in the 6H crystals.

The early optical studies of SiC have been reviewed in some detail by Choyke and Patrick [10] in 1973, and we will simply outline some of the important results concerning the cubic material. Choyke, Hamilton, and Patrick [7] studied the low temperature optical absorption edge in Lely-grown crystals of cubic SiC from which they inferred the indirect character of the optical transitions and a value of 2.390 eV for the exciton band gap. In this seminal paper they also reported a five-line PL spectrum comprising a 2.379 eV zero phonon line (ZPL) and its TA, LA, TO, and LO phonon replicas. On the basis of comparison with the PL spectra of N-doped 6H and 15R crystals grown in the same furnace run [16], the authors attributed the 5-line PL spectrum in the cubic crystals to the recombination of excitons bound at neutral N_C donors [7]. Two years later, Hartman and Dean [17] carried out an optical study of the symmetry of the recombination center giving rise to this 5-line PL spectrum. They stated that their work was partially motivated by the failure to observe a nitrogen-related spin resonance signal in an undoped crystal of cubic SiC which exhibited the supposed nitrogen bound exciton PL. Their thorough investigation of the Zeeman and uniaxial stress splitting of these sharp-line PL spectra determined that the luminescence signature does arise from the recombination of excitons bound to neutral donors, and that the donors are isolated point defects (i.e. not complexes). These results are consistent with but do not provide proof of the attribution to N_C donors.

In 1977, Dean, Choyke, and Patrick [8] studied the two electron satellite transitions associated with the "nitrogen" bound exciton spectrum. They measured the donor ground state-to-excited state separations from which a 53.6 +/-0.5 meV value of the donor binding energy was inferred. This highly accurate value of the binding energy provides a means of associating the donors giving rise to the bound exciton spectrum in cubic SiC with the donors involved in the donor-acceptor pair PL spectra which have been observed in both Al- and B-doped samples. Choyke and Patrick [9] first reported the observation of sharp close DAP PL spectra in Al-doped Lely-grown crystals of cubic SiC. They attributed these PL features to N-Al pairs, and these data were later used to infer [8,10] values of the donor and acceptor binding energies ($E_D(N)$ = 53 meV; $E_A(Al)$ = 257 meV). Similarly, KUWABARA, YAMADA, and TSUNEKAWA [18] inferred $E_D(N)$ = 55 meV and $E_A(B)$ = 735 meV from DAP PL spectra observed in B-doped cubic SiC. Although the identity of the residual donor involved in all of these PL phenomena cannot be determined with certainty, it can be concluded on the basis of the donor binding energies (E_D is ~53-55 meV in all cases) that the same residual donor is responsible for the bound exciton spectrum in undoped samples and the DAP spectra in Al- and B-doped samples of cubic SiC. In addition, the facts that the DAP PL in B-doped SiC is a type I spectrum [18] (donor and acceptor on the same sublattice) and that in Al-doped SiC is a type II spectrum [9] (donor and acceptor on opposite sublattices), and that EPR has shown B to be located on the carbon sublattice [12] are all consistent with the assumption that the residual donor is located on the carbon sublattice.

3. EPR Spectra in CVD-Grown Thin Film Cubic SiC

CARLOS and coworkers [4] have previously reported low-temperature EPR spectra of undoped CVD films of SiC. They observed an isotropic, well-resolved, 3-line spectrum characteristic of a central hyperfine interaction with a nucleus having spin I = 1. They pointed out that ^{14}N is the only principal isotope having a nuclear spin of 1 (the other two are ^{6}Li and deuterium) and therefore attributed the spectrum to nitrogen. The small, isotropic hyperfine interaction (A=1.2G) is consistent with expectations for a hydrogenic donor and is in excellent agreement with that reported by ALTAISKII et al.[15] for Lely-grown cubic SiC. The well-resolved 3-line spectrum was observed in a sample for which a $2X10^{17}cm^{-3}$ N concentration was measured from the EPR intensity. However, CARLOS et al. found that a sample having a N concentration about a factor of two larger did not exhibit a well-resolved hyperfine structure in its EPR spectrum. They attributed this absence of hyperfine structure to broadening due to exchange interactions. Because the g-value of this structureless spectrum was identical to the 2.0055 g-value measured for the 3-line N spectrum, it was also attributed to N donors. This positive g-shift (g>the 2.0023 free electron value) is relatively unusual for donors and this led these workers to suggest that it was unlikely that the structureless spectrum of the higher concentration sample consisted of the superposition of spectra from two distinct donor centers. An attempt to measure the donor binding energy from the temperature dependent quenching of the EPR spectrum was relatively unsuccessful because of severe changes in the line shape with increasing temperature.

Recently, we have carried out an analogous EPR study on several undoped CVD films of SiC which were grown with different values of the C/Si ratio in the source gas. Representative spectra are shown in Fig. 1 and the results can be summarized as follows. The EPR spectra of samples grown with high C/Si ratios (2.4 and 3.0) are

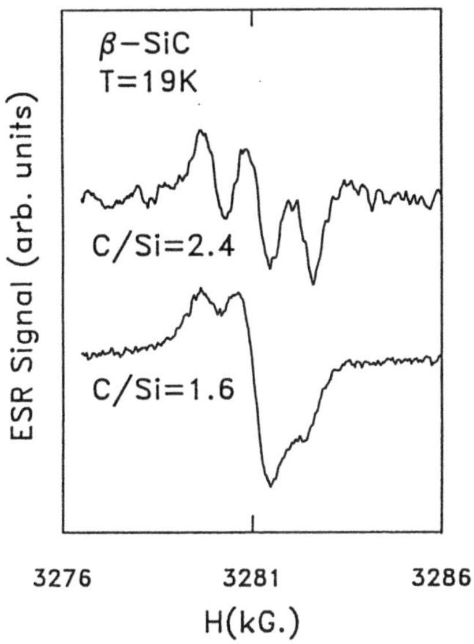

Fig.1 Low-temperature EPR spectra obtained at 9 GHz from CVD films of cubic SiC grown with C/Si source gas ratios of 1.6 and 2.4. Microwave power level was ~2 microwatts.

dominated by the well-resolved 3-line spectrum attributed to isolated N donors while those from samples grown with low C/Si ratios (1.2 and 1.6) can be interpreted as having a substantial contribution due to an underlying broad, structureless resonance with the same 2.0055 g-value.

Obviously this interpretation differs from that of the previous work of CARLOS et al. [4] and is motivated by a detailed line shape simulation. It is possible to fit the peaks in the observed line shape for low C/Si ratios with the 3-line spectrum alone only by assuming a broad distribution in the hyperfine interaction, A, and a decreasing $|A|$ with decreasing C/Si ratio. However, this procedure is not successful in fitting other portions of the lineshape, primarily the broad wings. Furthermore, as described in the next section, the PL data provide strong evidence for the existence of a ~54 meV donor in addition to the 15-20 meV shallow donor which dominates the electrical properties of the n-type undoped films. Thus an interpretation of the EPR spectra in terms of two independent spin resonance centers would be consistent with the results of other experimental measurements.

4. Photoluminescence Spectra in CVD-Grown Thin Film Cubic SiC

FREITAS et al.[5,6] have reported the observation of the characteristic 5-line bound exciton PL spectrum in undoped films of CVD-grown cubic SiC, an example of which is shown in Fig. 2. However, the quality of the PL spectra in the CVD films has not yet equalled that of spectra obtained from Lely crystals [7,17]. Even in films which are removed from their Si substrates and annealed at high temperature [5,6] exciton line widths are ~2 meV in the best spectra. This is to be compared with exciton line widths ~0.15 meV for published spectra from Lely crystals [17]. In addition, the intensity of the ZPL for the bound exciton spectra in CVD films is always much less than that of the phonon replicas (see Fig. 2) while all five lines (ZPL and four replicas) have comparable intensities in the PL spectra of Lely crystals [7,17]. Apparently, the limiting factors which cause these disparities in spectral quality are internal strain and high concentrations of charged centers in the

CVD films. Consequently, features such as the two electron satellites of the bound exciton spectrum have not yet been observed in the PL spectra of CVD films of cubic SiC. However, sharp-line close DAP PL spectra have recently been observed [4] in Al-doped films (Fig. 3). These spectra correspond in detail to those

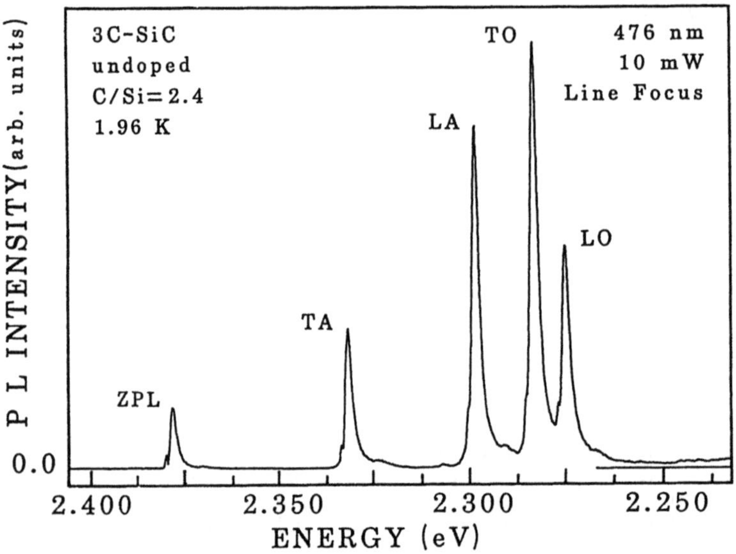

Fig.2 Donor bound exciton PL spectrum obtained at 1.96 K from an undoped CVD-grown film of cubic SiC. The ZPL line width is ~2 meV and the intensity of the ZPL relative to those of the phonon replicas is representative of the best yet observed in CVD films.

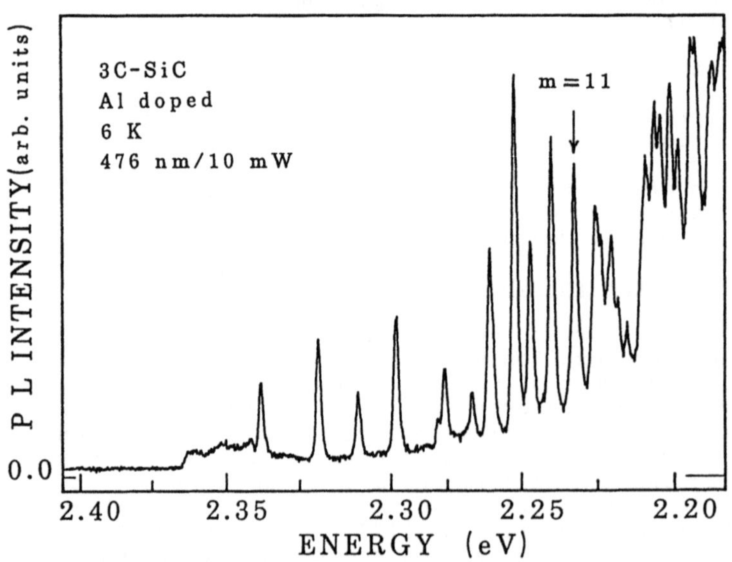

Fig.3 Low temperature photoluminescence spectra from an Al-doped cubic SiC film showing sharp-line close DAP spectra. The arrow indicates the 2.233 eV PL band whose temperature dependence is plotted in Fig. 4. This band corresponds to a pair separation of 1 nm (m=11).

reported previously for Al-doped Lely crystals [9] and quantitative analysis of the energies of the close pair transitions has confirmed [4] the values of the donor and acceptor binding energies ($E_D(N)$ = 53 meV; $E_A(Al)$ = 257 meV).

On the basis of the independently measured 53-55 meV values of the binding energies, it appears that the same residual donor which gives rise to the five-line bound exciton spectrum is also involved in the sharp-line close DAP PL spectra observed in the CVD films of cubic SiC. The obvious question which arises is whether or not this residual donor with its optically measured ~54 meV binding energy is related to the 15-20 meV donor which dominates the electrical properties of the n-type films. It has been observed [19] that optical transitions which depend only upon impurity ground state energies can remain sharp and unshifted at relatively high impurity concentrations for which thermal activation or binding energies for carriers bound to the impurities can be depressed by as much as 10-40% because of apparent overlap in the impurity excited states. In addition, SEGALL et al. [2,3] have invoked a concentration dependent reduction in the thermal activation energy to explain the variation in donor binding energies inferred from temperature dependent Hall effect measurements on n-type CVD films of cubic SiC.

In order to investigate the possibility that the donor binding energy as measured by a thermal activation technique (such as temperature dependent Hall measurements) might be reduced relative to that measured optically, we have studied the thermally activated quenching of the DAP PL intensity. The optically measured 54 meV binding energy of the donors involved in the DAP PL is much smaller than that of the Al acceptors (257 meV); this means that the quenching of the DAP PL with increasing temperature should be controlled by the ionization of the donors only and a thermal activation energy can be obtained from the quenching curve [19]. The temperature dependence of the intensity of the 2.233 eV close DAP PL line is shown in Fig. 4. This particular PL line, designated by an arrow in Fig. 3, corresponds to a pair separation of 1 nm. The slope of the high temperature exponential decay portion of the thermal quenching curve yields an activation energy of 53.5 meV. This result indicates that for the donors participating in the DAP

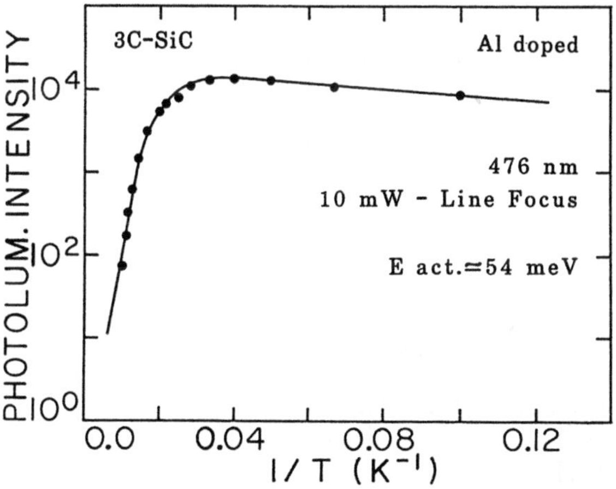

Fig.4 Photoluminescence intensity as a function of reciprocal temperature for the 2.233 eV N-Al DAP line (m=11, pair separation of 1 nm) in an Al-doped film of cubic SiC.

PL process there is no significant difference between the optically determined binding energy and the thermal activation energy. Therefore, this ~54 meV donor is distinctly different from the 15-20 meV donor which is detected in the temperature dependent Hall measurements on n-type films.

5. Optically Detected Magnetic Resonance in Al-Doped CVD SiC

In the optically detected magnetic resonance (ODMR) technique [20], the paramagnetic resonance spectrum is detected as a change in the efficiency of a PL band. The pertinent band is selected by appropriate optical filters and its intensity is monitored as the sample is subjected to microwave radiation and high intensity magnetic field dictated by the resonance condition. If the recombination mechanism is directly or indirectly spin dependent, the paramagnetic resonance spectrum can be detected as a change in the intensity of the selected PL band. Recently we have initiated ODMR studies of CVD films of cubic SiC and we present here some preliminary results which are pertinent to the identification of the residual 54 meV donor.

ODMR spectra have been obtained from Al-doped CVD films of SiC which exhibit the residual donor-Al pair PL spectra. Figure 5 shows the ODMR spectrum obtained from the same sample whose DAP PL spectra are depicted in Fig. 3. The spectrum was recorded on the integrated intensity of the DAP PL bands at a microwave frequency of 24 GHz. On the basis of comparison of the spectrum with the EPR [4,12-15] and ODMR [21] studies of cubic and hexagonal SiC the resonance is attributed to a donor. In the ODMR spectra for 4H and 6H SiC a three-line hyperfine structure was resolved [21] which identified N as the donor involved in the DAP PL. However, no hyperfine splitting has been resolved in the ODMR of the cubic SiC. Because of the high donor and acceptor concentrations, the average separation of a pair is small. Apparently, exchange interactions at these small separations broaden the ODMR spectra so that hyperfine interactions (which are at least 10 times smaller in cubic material) are not resolved. The absence of a hyperfine interaction precludes a definitive identification of the donor participating in the DAP PL

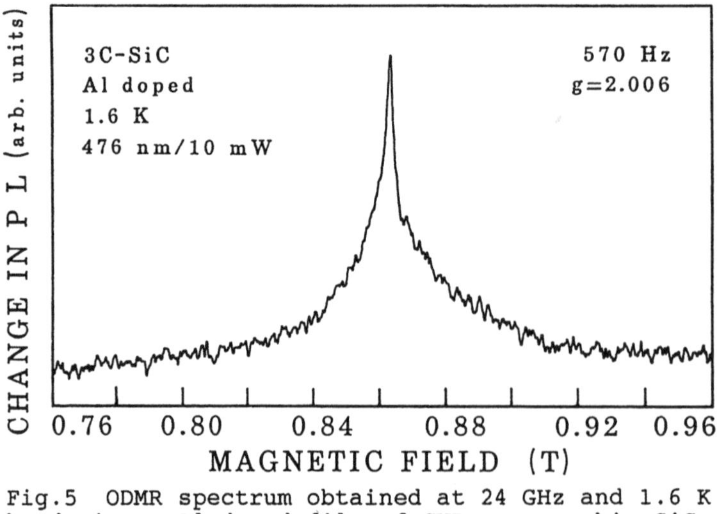

Fig.5 ODMR spectrum obtained at 24 GHz and 1.6 K from the DAP PL bands in an Al-doped film of CVD-grown cubic SiC. The modulation frequency for the microwaves was 570 Hz.

and giving rise to the ODMR signal, but the 2.006 g-value is obviously consistent with the 2.0055 value obtained for the nitrogen donor EPR in the CVD films (Section 3). It is possible that ODMR spectra obtained from samples having lower donor and Al concentrations will exhibit a hyperfine splitting and thereby provide a more reliable identification of the residual donor.

6. Summary and Conclusions

On the basis of the above discussion it can be concluded that there exists in Lely- and CVD-grown cubic SiC a residual donor with binding energy of about 54 meV. Values of the binding energy of this donor which are close to 54 meV have been inferred independently from the transition energies of donor-Al [4,9] and donor-B [18] close pair PL spectra and from the thermal quenching curve for the intensity of these pair spectra in Al-doped films. In addition, a five line neutral donor bound exciton spectrum is observed in these films [5,6] which is identical to that observed in Lely crystals [7]. The Zeeman and uniaxial stress studies [17] of this bound exciton PL in Lely crystals indicate that the spectrum arises from a neutral, isolated (point defect) donor. A highly accurate value of 53.5 meV for the donor binding energy in the Lely crystals was determined [8] from the two electron satellites of this donor bound exciton spectrum. Thus it appears that the same donor which participates in the DAP PL process in Al-doped material also gives rise to the bound exciton PL spectrum in n-type material. Because of the sharp bound exciton and close DAP PL spectra which it exhibits, it can be concluded that the 54 meV donor exists in a high quality crystalline environment in the SiC films.

On the strength of these PL results, and the fact that well-resolved 3-line EPR spectra characteristic of isolated, neutral N donors have been observed in CVD films, and the observation of a donor resonance in the ODMR spectrum of the donor-Al pair PL which has a g-value consistent with the ^{14}N EPR, it would appear that N is the most probable attribution for the residual 54 meV donor. It will be recalled that the bound exciton spectrum and the DAP spectra were also attributed to N donors in the earlier PL studies of Lely crystals[7-10]. Furthermore, the equivalence of the optically measured donor binding energy and the thermal activation energy for the donor indicates that the concentration of this 54 meV residual donor is not large enough to cause a concentration dependence of the activation energy ($N_D \leq 10^{17} cm^{-3}$) [19]. Therefore, the 15-20 meV donor observed in Hall effect measurements is independent of the 54 meV donor which we attribute to isolated N on the carbon sublattice. If the 15-20 meV donor is related to N, it must be in association with other defects (i.e. it is not isolated) or it must occur in inhomogeneities in the films where the N impurities are present at much higher concentrations and/or in highly distorted bonding configurations.

Acknowledgement

This work has been partially supported by an ONR contract.

References

1. A. Suzuki, A. Uemoto, M. Shigeta, K. Furukawa, and S. Nakajima: Appl. Phys. Lett. 49, 450 (1986)

2. B. Segall, S.A. Alterovitz, E.J. Haugland, and L.G. Matus: Appl. Phys. Lett. 49, 584 (1986)

3. B. Segall, S.A. Alterovitz, E.J. Haugland, and L.G. Matus: Appl. Phys. Lett. 50, 1533 (1987)

4. W.E. Carlos, W.J. Moore, P.G. Siebenmann, J.A. Freitas, Jr., R. Kaplan, S.G. Bishop, P.E.R. Nordquist, Jr., M. Kong, and R.F. Davis: in Proc. Mats. Res. Soc. Symp., Anaheim, CA, Apr 1987, in press

5. J.A. Freitas, Jr., S.G. Bishop, A. Addamiano, P.H. Klein, H.J. Kim, and R.F. Davis: in Mats. Res. Soc. Symp. Proc., Vol. 46, 581 (1985)

6. J.A. Freitas, Jr., S.G. Bishop, J.A. Edmond, J. Ryu, and R.F. Davis: J. Appl. Phys. 61, 2011 (1987)

7. W.J. Choyke, D.R. Hamilton, and L. Patrick: Phys. Rev. 133, A1163 (1964)

8. P.J. Dean, W.J. Choyke, and L. Patrick: J. Lum. 15, 299 (1977)

9. W.J. Choyke and L. Patrick: Phys. Rev.B 2, 4959 (1970)

10. W.J. Choyke and L. Patrick: in Silicon Carbide-1973, ed. by R.C. Marshall, J.W. Faust, Jr., and C.E. Ryan (University of South Carolina, Columbia, SC 1974) p.261

11. G. Lancaster: Electron Spin Resonance in Semiconductors (Plenum, New York 1967) p.16

12. H.H. Woodbury and G.W. Ludwig: Phys. Rev. 124, 1083 (1961)

13. A.I. Veinger: Sov. Phys.-Semic. 1, 14 (1967)

14. I.M. Zaritskii, L.A. Shulman, and I.N. Geifman: Sov. Phys.-Solid State 11, 22 (1969)

15. Yu.M. Altaiskii, I.M. Zaritskii, V.Ya. Zevin, and A.A. Konchits: Sov. Phys.-Solid State 12, 2453 (1971)

16. W.J. Choyke and L. Patrick: Phys. Rev. 127, 1868 (1962)

17. R.L. Hartman and P.J. Dean: Phys. Rev.B 2, 951 (1970)

18. H. Kuwabara, S. Yamada, and S. Tsunekawa: J. Lum. 12/13, 531 (1976)

19. P.J. Dean: in Progress in Solid State Chemistry, Vol. 8 (ed. by J.O. McCaldin and Y.G. Somarjai (Pergamon, Oxford 1973) p.1

20. J.J. Davies: J. Cryst. Growth 72, 317 (1985)

21. N.G. Romanov, V.A. Vetrov, and P.G. Baranov: Sov. Phys.-Solid State 20, 96 (1986)

SiC Production Chemistry from Olefinic Hydrocarbons on Si(100)

W.J. Choyke, R. Wallace, L. Henderson, P. Taylor, M.J. Dresser, and J.T. Yates, Jr.

Surface Science Center, Department of Chemistry and Department of Physics, University of Pittsburgh, Pittsburgh, PA 15260, USA

We have utilized molecular beam methods, temperature programmed desorption, and Auger spectroscopy to study the efficiency of SiC production on Si(100). It has been found that the C=C double bond is the key functional group for interaction with Si dangling bond sites on Si(100) at 100K. This C=C bond is broken forming an anchored free radical hydrocarbon species. C-C single bonds are inactive for this initial attachment of hydrocarbon to Si sites.

Two olefins have been compared in their interaction with Si(100). Both ethylene, $H_2C{=}CH_2$, and propylene, $H_3CCH{=}CH_2$ are efficiently captured by dangling bond Si sites, producing anchored free radical surface species. Surprisingly, however, ethylene-derived species are liberated at ~500 K as ethylene gas (probability = 1) while propylene-derived species on the surface are efficiently thermally converted to SiC with a probability of about 2/3. The remainder of the propylene is liberated as propylene gas.

This observation is explained by an enhanced reactivity of methyl-C-H bonds in the anchored propylene-derived radical surface species which can be justified on thermodynamic grounds. It is this second reaction channel which controls the deposition rate of C on Si (100), and hence the efficiency of SiC formation.

Work supported by the Office of Naval Research

Surface Structures of ß-SiC, 6H-SiC and Pseudomorphic Si Adlayers

R. Kaplan

Naval Research Laboratory, Washington, DC 20375, USA

Introduction

A systematic understanding of the structure and composition of SiC surfaces on an atomic scale has begun to emerge during the last few years. As in the case of compound semiconductors generally, free SiC surfaces in vacuum can exhibit a range of stable configurations or reconstructions, each with its own characteristic stoichiometry and structure. Earlier studies have examined SiC surfaces cleaned simply by annealing in ultra-high vacuum (UHV)[1–3], by Ar^+ ion bombardment of samples at elevated temperature[4], or by surface oxide reduction by exposure of heated samples to Ga or As beams[5]. In all of these studies, phase transformations on clean surfaces were produced by annealing in vacuum at successively higher temperature. Since annealing leads to loss of surface Si[6], the transformations to Si-poor configurations were irreversible. Furthermore, Si-rich surfaces could not reliably be obtained. In spite of these limitations, most of the structures to be discussed here have previously been observed. This paper describes a new method for obtaining all of the stable phases of a given SiC surface in a reversible and reproducible fashion. A new phase of (001) β-SiC has been found, the c (4×2), which appears to be the intrinsic structure of the Si-terminated surface. Additional observations relating surface composition and structure are presented for (001) and (111) β-SiC, and (0001) 6H-SiC, and models for several of the structures are discussed.

Experimental

Low energy electron diffraction (LEED), Auger (AES), and energy loss (ELS) measurements were made on β-SiC in the form of nominally 10 μm thick films grown on Si substrates, and on 6H samples comprising thin, free-standing platelets. The former were provided by L.G. Matus, NASA-Lewis Research Center, and R.F. Davis and coworkers, N. Carolina State University, the latter by A. Addamiano of this laboratory. Clean, ordered surfaces were achieved by heating samples near 850°C in the presence of a Si flux of about 10^{14} cm^{-2} sec^{-1}, obtained from a resistively heated Si slab. In this manner a volatile surface oxide was formed on and evaporated from the SiC in a controlled fashion.

Springer Proceedings in Physics, Vol. 34 Amorphous and Crystalline Silicon Carbide
Editors: G.L. Harris and C.Y.-W. Yang

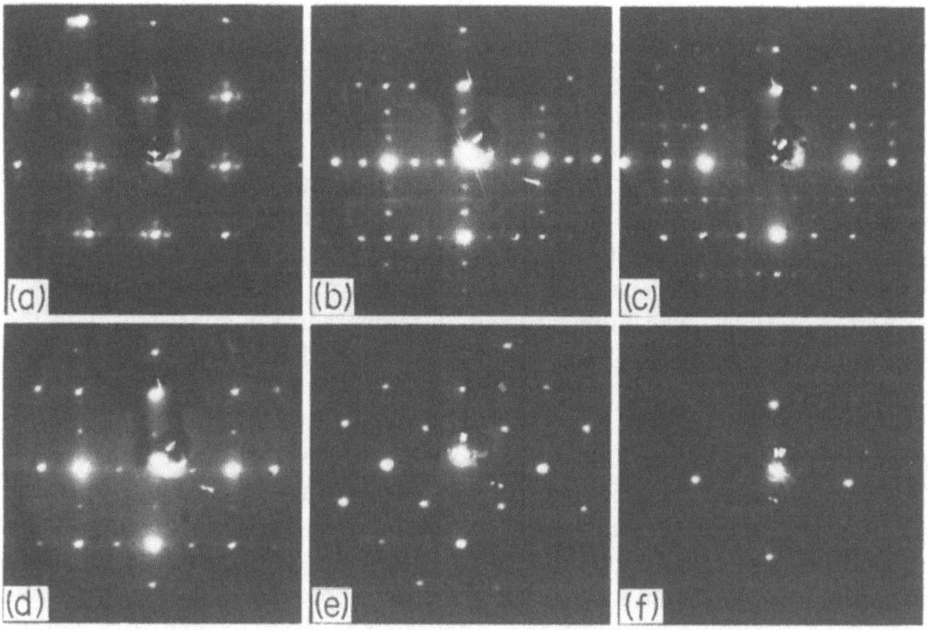

Fig. 1 — LEED patterns from (001) β-SiC at room temperature, 90 eV. (a) with ordered partial overlayer of oxygen. (b) 3 × 2. (c) *c* (4 × 2). (d) 2 × 1. (e) *c* (2 × 2). (f) 1 × 1. The 5 clean surface phases are in order of decreasing Si content.

Incomplete oxide removal yielded an ordered oxygen overlayer as indicated in Fig. 1a for (001) β-SiC. After cleaning, the surface composition and structure subsequently could be altered by annealing in greater or lesser Si flux or in vacuum; the resulting phase transformations were reversible. Surface phases were characterized by their LEED geometry, AES intensity ratio Si^{LVV}/C^{KLL}, the shapes of the Si^{LVV} and C^{KLL} peaks, the value of the plasmon energy $\hbar\omega_p$ and its dependence on primary energy E_p, and the presence of interband loss peaks characteristic of sp^3 tetrahedral bonding.

β-SiC (001)

The (001) surfaces of smooth, highly reflecting β-SiC samples yielded 2-domain c (4 × 2) LEED patterns after initial cleaning, as shown in Fig. 1c, and Si^{LVV}/C^{KLL} ~4. Comparably treated samples with rougher surfaces exhibited 2-domain 2 × 1 patterns, Fig. 1d, i.e. the 1/4th order spots were absent. The AES and ELS results from both types of sample were identical, and implied a SiC surface terminated in a single Si atom layer. Thus, the appearance of 1/4th order spots indicates a long range ordering in the outer Si atom layer, facilitated by good surface quality. This behavior recalls that of Si (001), which exhibits a c (4 × 2) phase[7] that partially disorders at temperatures above 200 K. The Si c (4 × 2) is believed[8,9] to originate in an ordered configuration of the buckled dimers responsible for the basic 2 × 1 reconstruction. This configuration is "antiferromagnetic" (AF), in the sense that all neighboring dimers are oppositely oriented, as shown in Fig. 2a. Above 200 K, the 1/4th order spots disappear from the Si LEED pattern, due to thermal disordering of the buckled dimer configuration. This order-disorder transition apparently occurs above room temperature for the SiC c (4 × 2), assuming that the nature of the reconstruction is the same for both materials. The higher transition temperature is consistent with the smaller Si atom spacing on SiC relative to Si, which should lead to a stronger dimer-dimer interaction for SiC surfaces.

After suitable exposure of the (001) β-SiC to a Si flux at about 700°C, the value of Si^{LVV}/C^{KLL} increased to ~5, and the c (4 × 2) was replaced by a pattern, shown in Fig. 1b, previously designated[1] 2-domain 3 × 2. Each domain of this pattern contained both sharp spots and streaks, the latter implying disorder in the direction normal to the dimer rows. The patterns' appearance did not vary with sample selection or surface preparation. According to the AES and ELS results, the 3 × 2 structure must contain additional Si adsorbed on the c (4 × 2) Si-terminated surface. Prominent among these results are a split Si^{LVV} peak, which implies the presence of both Si-Si and Si-C bonds, and an additional plasmon peak near the characteristic Si energy, which grows in strength with decreasing E_p. The 20% smaller atom separation on SiC relative to Si precludes the possibility of a complete Si adlayer with atom density equal to that of the substrate. In the most likely adlayer structure,[1] every third row of Si atoms parallel to the dimer direction is missing, in each of the two domains. This leads to well-defined 1/3rd order spots as observed. If the dimers were symmetric, they would be indistinguishable, and a sharp 3 × 2 diffraction would result. Thus the streaking of the 1/2 order features implies that the dimers are buckled as for (001) Si, and have either an incipient ferromagnetic (F) or layered antiferromagnetic (LAF) configuration at room temperature. In the LAF arrangement, dimers in a given row have like orientation, but adjacent rows are oppositely oriented. This arrangement is illustrated in Fig. 2b. The configuration cannot be AF as for Si (001) and bare SiC (001), since 1/4th order LEED features were absent. The sharpness of the streaks along the dimer directions indicates that dimers in a given row are well ordered. Examination of the structure factor for diffraction normal to the rows suggests that the dimers are ordered within each pair of rows bounded by vacancy rows. However, coupling across the latter is much weaker, so that long range ordering normal to the rows is absent. A comparable streaking was observed for the c (4 × 2) pattern in (001) Ge[10]: ordering occurred first for dimers within individual rows, and only at lower temperatures did rows

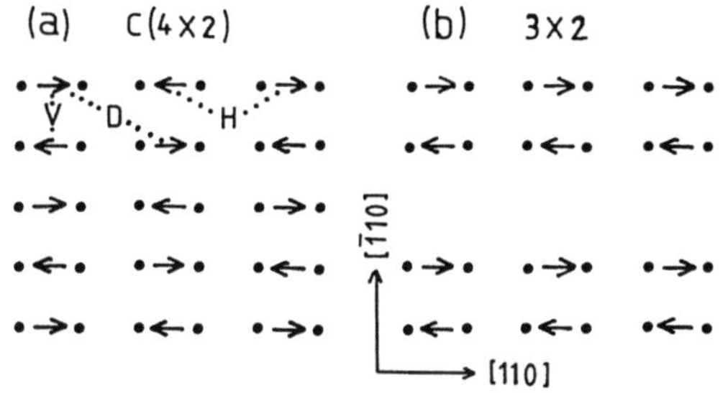

Fig. 2 — Buckled dimer configurations for (001) reconstructions. (a) *c* (4 × 2). (b) 3 × 2. The dots indicate surface atom positions before dimer formation. Single domain shown.

become correlated. Extrapolation from calculations[8] of the coupling strengths of different dimer pairs suggests that LAF ordering dominates on the (001) SiC 3 × 2 surface.

Annealing of the (001) surface in vacuum at about 950-1000 °C yielded Si^{LVV}/C^{KLL} ~3, and LEED patterns with c (2 × 2) geometry, as in Fig. 1e. AES and ELS results implied a surface terminated in substantially less than one monolayer of Si, and with altered Si-C bonding. A model proposed earlier[1] for this reconstruction assumes termination in a full Si monolayer, the dimers on alternating rows being displaced along the row one atom separation, thus directly yielding a c (2 × 2) arrangement of dimers as shown in Fig. 3a. This arrangement is in fact the basis for the "c (2 × 2) family" of reconstructions found theoretically[8,9] to have a substantially higher energy than those of the "2 × 1 family," and not previously observed. This fact, plus the AES and ELS results cited above, particularly the surface Si concentration, are not consistent with the proposed model. An alternative model under consideration, Fig. 3b, places 1/2 monolayer of Si atoms in the plane of the surface C, at the centers of alternating squares of C atoms. The Si bonding configuration is assumed to be planar sp^2d. It is possible that the energy gain due to $p-d$ promotion and a slightly larger than ideal bond length is compensated by the elimination of *all* dangling bonds.

Additional annealing in vacuum led to further reduction in Si^{LVV}/C^{KLL}, and a 1 × 1 LEED pattern as shown in Fig. 1f. The presence of surface graphite could just be detected in the AES and ELS spectra. The 1 × 1 pattern may be representative of a C-terminated (001) sample, but the surface was probably inhomogeneous; certainly the AES and ELS results indicated a departure from normal bonding, averaged over the surface. Continued annealing led eventually to graphitization, clearly evident in the LEED patterns.

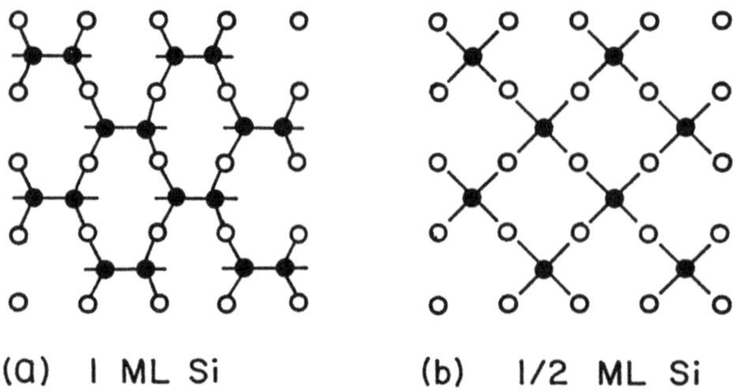

Fig. 3 — Possible structures for (001) $c\ (2 \times 2)$ phase. (a) full Si monolayer (Ref 1). (b) 1/2 Si monolayer, planar sp^2d Si bonding.

β-SiC (111) and 6H-SiC (0001)

The Si-terminated β-SiC (111) and 6H-SiC (0001) surfaces gave essentially equivalent LEED, AES and ELS results, as expected from the crystal structure. Three distinct surface phases were observed, as shown in Fig. 4: 3×3, $\sqrt{3} \times \sqrt{3}$ R30° and 1×1, with decreasing values of Si^{LVV}/C^{KLL} ~5.5, 3 and 2 respectively. The 3×3 phase could be obtained by the same procedure used for the (001) 3×2, namely, exposure at about 700°C to a Si flux. As with the 3×2 phase, a split Si^{LVV} AES peak and an additional plasmon loss peak most evident for small E_p, indicated the presence of Si-Si bonding, i.e. Si adsorbed on a Si-terminated surface. This 3×3 phase is believed to be the $N = 1$ member of the $(2N + 1) \times (2N + 1)$ class of reconstructions[11], whose prototype is the Si 7×7 DAS structure[12]. Figure 5 shows a perspective view of a single unit cell of the proposed structure. It consists of a partial double layer of Si atoms on a Si-terminated surface, with adatoms at 3-fold top sites, and dimers and vacancies as indicated. Solid and open circles simply represent alternating layers of atoms. The lattice distortion requisite to dimer formation (double lines in figure) is much smaller than for Si (111), due to the smaller atom separation on SiC. The model assumes normal Si-Si and Si-C bond lengths, and requires relatively minor bond angle distortions. It is noteworthy that a $3\times$ vacancy periodicity serves to accommodate the Si/SiC lattice mismatch for both the (001) 3×2 and (111) or (0001) 3×3 structures.

The $\sqrt{3} \times \sqrt{3}$ and 1×1 phases which appear as the surface is progressively depleted of Si have not as yet been studied in detail, and no structures are proposed for them at this time. As in the case of (001) surfaces, Si depletion can lead to graphitization and surface inhomogenity, and care must be taken to minimize these effects.

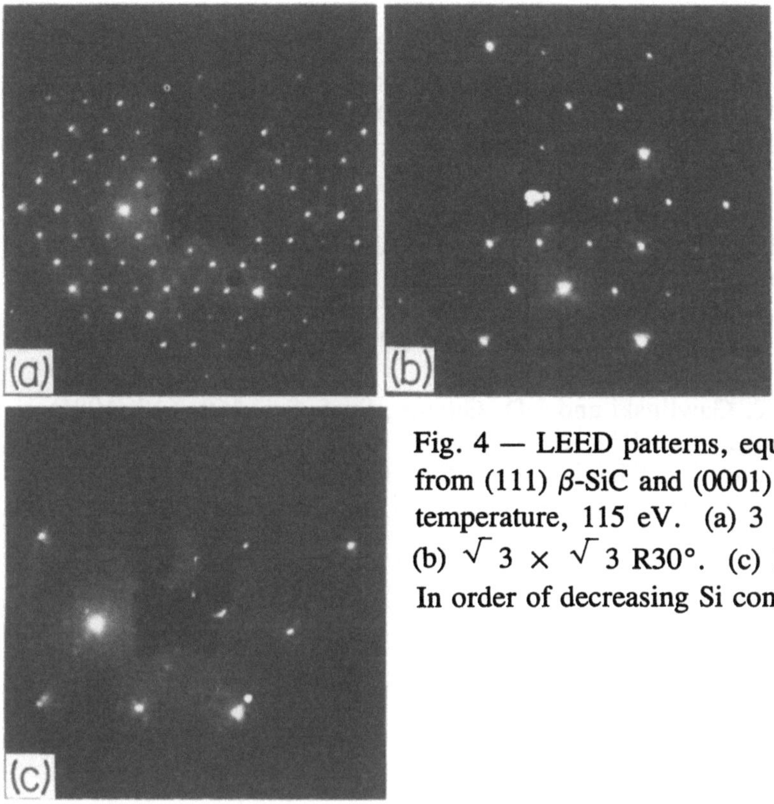

Fig. 4 — LEED patterns, equivalent results from (111) β-SiC and (0001) 6H-SiC at room temperature, 115 eV. (a) 3 × 3. (b) $\sqrt{3} \times \sqrt{3}$ R30°. (c) 1 × 1. In order of decreasing Si content.

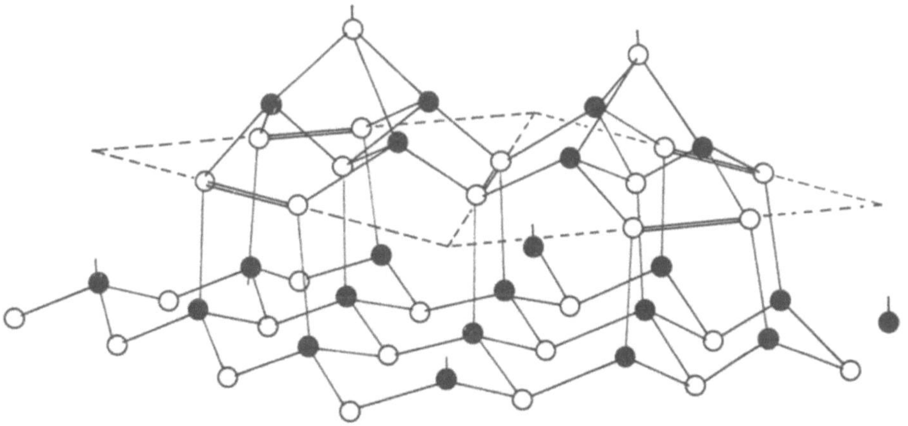

Fig. 5 — Perspective view of the (2N + 1) × (2N + 1) model structure for the 3 × 3 phase of (111) β-SiC and (0001) 6H-SiC surfaces. Open and filled circles are used solely to identify atoms on alternating layers. The uppermost double layer (Si-terminated) of the SiC is shown, with a single unit cell of the 3 × 3 adsorbed Si structure indicated by the dashed lines.

References

(1) M. Dayan, J. Vac. Sci. Technol. A**4**, 38 (1986).
(2) M. Dayan, J. Vac. Sci. Technol. A**3**, 361 (1985).
(3) A.J. van Bommel, J.E. Crombeen and A. van Tooren, Surf. Sci. **48**, 463 (1975).
(4) R. Kaplan, J. Appl. Physics **56**, 1636 (1984).
(5) R. Kaplan and T. M. Parrill, Surf. Sci. **165**, L45 (1986).
(6) J.J. Bellina, Jr. and M.V. Zeller, Appl. Surf. Sci. **25**, 380 (1986).
(7) T. Tabata, T. Aruga and Y. Murata, Surf. Sci. **179**, L63 (1987).
(8) J. Ihm, D.H. Lee, J.D. Joannopoulos and J.J. Xiong, Phys. Rev. Lett. **51**, 1872 (1983).
(9) A. Saxena, E.T. Gawlinski and J.D. Gunton, Surf. Sci. **160**, 618 (1985).
(10) S.D. Kevin, Phys. Rev. B**32**, 2344 (1985).
(11) Guo-Xin Qian and D.J. Chadi, J. Vac. Sci. Technol. B**4**, 1079 (1986).
(12) K. Takayanagi, Y. Tanishiro, S. Takahashi and M. Takahashi, Surf. Sci. **164**, 367 (1985).

Study of the Origin of Residual Carriers in CVD-Grown 3C-SiC by Photoluminescence and Electron Spin Resonance

H. Okumura, M. Yamanaka, M. Shinohara*, E. Muneyama**, S. Kuroda, K. Endo, I. Nashiyama, H. Daimon***, E. Sakuma, S. Misawa, and S. Yoshida*

Electrotechnical Laboratory,
1-1-4, Umezono, Tsukuba, Ibaraki 305, Japan

Abstract PL and ESR spectra were measured for undoped and N-doped 3C-SiC epitaxially grown by CVD, in order to investigate the origin of the residual carriers in the SiC films. The PL spectra of lightly N-doped samples show broad peaks besides the sharp peaks observed for undoped ones. The spin densities of undoped samples at 4.2K are estimated to be much lower than the carrier densities at 300K. These results suggest that the main origin of the residual carriers in undoped 3C-SiC is not the donor associated with nitrogen.

1. Introduction

Since the chemical vapor deposition (CVD) technique has been developed to obtain 3C-SiC film epitaxially grown on Si,[1] several attempts have been made to fabricate electronic devices using 3C-SiC epilayers grown by the CVD technique. High quality epilayers are necessary for the fabrication of electronic devices. However, the carrier densities of the undoped SiC epilayers are not low enough for device fabrication, because of the existence of residual carriers with the density of $10^{16}/cm^3$. It is required to clarify the origin of such residual carriers and to reduce the amount of them. In the past, Choyke *et al.*[2,3] reported the origin to be nitrogen. However, we have recently reported the donor-density dependence of the activation energies of donors in N-doped samples, which suggests that the donors in undoped 3C-SiC epilayers may not be nitrogen impurities.[4] In this paper, we report the photoluminescence (PL) and electron spin resonance (ESR) measurements of undoped and N-doped 3C-SiC epilayers grown by CVD, and discuss the origin of residual carriers in undoped ones, based on the spectral differences.

2. Experiments

3C-SiC epilayers used in our measurements were grown on Si (001) substrates at 1350°C by CVD. SiH_4 and C_3H_8 were used as source gases in a flow of H_2 carrier gas. The details of the epitaxial growth were described elsewhere.[5] For the doping of nitrogen, N_2 gas diluted with H_2 was added to the reaction gases. The amount of N introduced to the epilayers was controlled by the flow rate of N_2 gas. The carrier density and Hall mobility of the samples were determined using van der Pauw method. The carrier density of N-doped samples increased with the increase of the N/Si atomic ratio in the reaction gases.

---

* On leave of absence from Central Eng. Lab., Nissan Motor Co., Ltd.

** Present address, UBE Scientific Analysis Laboratory.

*** Present address, Ube Lab., Ube industries Ltd.

PL spectra were measured at 4.2K by using a 488nm line from an Ar^+ laser. This excitation light was focused on the sample surfaces; the power at the sample surfaces was around $100W/cm^2$. Spectra were measured for central homogeneous portions of epilayers. Since the absorption coefficient of 3C-SiC for 488nm light is so small, the intensities of the measured PL spectra of samples with different thicknesses (14-28μm) were calibrated according to their respective thickness.

ESR spectra were measured using an X-band spectrometer in the temperature range of 4.2-300K. Prior to the measurements, Si substrates were removed with an $HF\text{-}HNO_3$ solution. Some pieces of the SiC thin films with the size of typically 3mm×10mm sealed up in a quartz sample tube with He gas were used for measurements.

3. Results and Discussion

Figure 1 shows the PL spectra of undoped and N-doped 3C-SiC epilayers for the spectral region of 2230-2350meV. Spectrum 1 in the figure is for the undoped epilayer having a carrier density of $2.4\times10^{16}/cm^3$ and a Hall mobility of $729cm^2/(V\bullet s)$ at 300K. Spectrum 1 exhibits four well-resolved emissions. The intensities of these peaks are stronger for high-density carriers and low-mobility epilayers.[6)]

Choyke *et al.*[3)] measured the PL spectra of undoped 3C-SiC and assigned the peaks in this region to the nitrogen bound exciton emission and its phonon replicas. Since then, it has been believed that the donors in undoped 3C-SiC are residual nitrogen atoms. The energies of the peaks observed in our measurements are shown in Fig.1, and are essentially the same as Choyke's results. These peaks are certainly interpreted as the phonon replicas taking account of the phonon energies at point X. They are considered to be due to the excitons bound to some donors located at point X with the exciton binding energy of around 15meV. However, the ground justifying the identification of these donors as N impurities has not been explicitly given in ref.3. Moreover, the purity and quality of the samples in Choyke's studies were not fully specified. Meanwhile, Freitas *et al.*[7)] reported the PL study of 3C-SiC epilayers grown by CVD. In their study, the PL spectrum of undoped epilayers showed peaks in the region of 2230-2350meV also, and that of the Al-doped one showed strongly red-shifted and broadened peaks in the same region. They also assigned the peaks observed for both the samples to the nitrogen bound exciton emission and its phonon replicas. The reason for such assignment is not clear.

If the emissions of undoped samples are due to nitrogen impurities, such impurities should come from the reaction gas during the CVD growth spontaneously. Adding N_2 to the reaction gases means only the increase of N_2 partial pressure. The intensities of these peaks might be expected to increase with intentional nitrogen doping. However, the intensities of these emissions do not increase with the amount of nitrogen introduced. Besides the peaks observed for undoped samples, other new peaks are observed and the intensities of these new peaks increase with the N/Si atomic ratio in the reaction gases, as shown in Fig.1. Because the series of the new peaks consist of

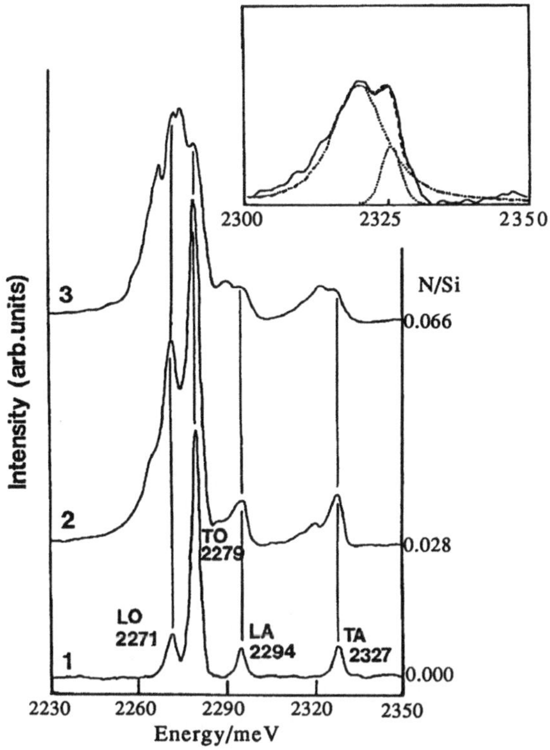

Figure 1. PL spectra of 1) undoped 3C-SiC epilayer and 2), 3) N-doped ones. The N/Si atomic ratio in the reaction gases for each sample is shown in the figure. The carrier concentrations at 300K for samples 1), 2) and 3) are $2.4\times10^{16}/cm^3$, $2.3\times10^{17}/cm^3$ and $3.6\times10^{17}/cm^3$, respectively. In the inset, the result of the deconvolution for TA phonon assisted emission of sample 3) is shown. The dashed line and dotted lines indicate the fitting curve which was obtained and its components, respectively.

four emission lines and the energy differences of the emissions correspond to those of the phonon energies at point X, they are interpreted as phonon replicas, as in the case of the emission lines of undoped samples. The energies of the new peaks are about 8meV lower than those of undoped epilayers. The new peaks of N-doped samples exhibit quite broad spectral widths, compared with those observed for undoped ones. Besides, in the spectra of undoped epilayers, no peaks are displayed at the energy positions corresponding to the peaks caused by doping nitrogen. Therefore, it is indicated that the peaks observed for undoped epilayers are not due to N_2 residual gases during the growth, and that residual nitrogen in the reaction gases is not the main origin of residual carriers.

In Fig.2, the ESR spectra of an undoped 3C-SiC epilayer and its temperature dependence are shown. These ESR signals consist of a sharp triplet structure and another broad one; the g-value was found to be 2.004. The triplet structure is due to the hyperfine interaction. The width of the hyperfine

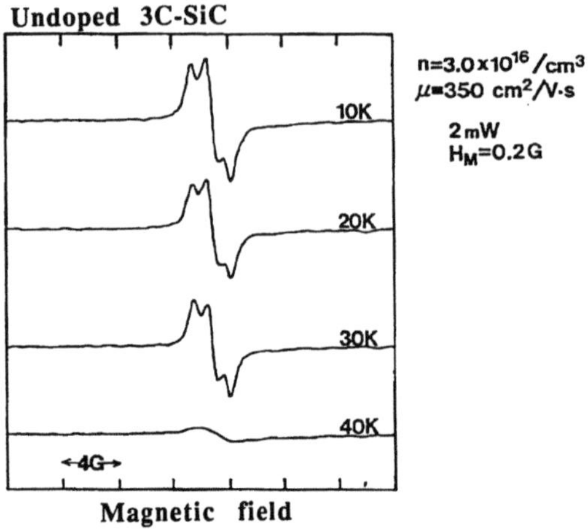

Figure 2. ESR spectra of an undoped 3C-SiC epilayer at various temperatures. *g*-value is 2.004. The signals consist of a triplet and a broad singlet structure.

splitting is estimated to be 1G. The anisotropy of ESR signals against the magnetic field was not observed. The ESR signal intensity decreases as the temperature is increased, and the triplet structure disappears above 40K.

The triplet structure of the hyperfine interaction is peculiar to the interaction between electron spins and nuclei with the spin of 1. Considering the growth process, the most plausible nucleus which brings about the triplet structure of the SiC samples is ^{14}N, the major isotope of nitrogen. In fact, Woodbury *et al.* found that the ESR triplet structure spectra of 6H-SiC single crystals result from the interaction with ^{14}N nuclei.[8] This result seemingly supports the belief that the origin of residual carriers in SiC is assigned to nitrogen impurities. However, the samples they measured were 6H polytype. The amount of the hyperfine splitting they observed was 12G, which is much larger than that for 3C-SiC. Altaiskii *et al.* and Carlos *et al.* also reported ESR spectra with 1G hyperfine splitting for 3C-SiC.[9,10] The only difference between 6H- and 3C-SiC is the stacking order of atomic planes. This large difference of hyperfine splitting cannot be explained assuming the same kind of ESR centers. The growth processes of 6H- and 3C-SiC are quite different. It is likely that the behavior of N atoms in 3C-SiC is different from that in 6H-SiC. We think that the triplet structure observed for the 3C-SiC shows only the existence of nitrogen atoms, but does not necessarily justify the assumption that the origin of residual carriers is nitrogen.

The intensity of the triplet structure observed for undoped 3C-SiC decreases with increasing temperature, although a tendency to saturate is observed at lower temperatures. This behavior is considered to be the result of the activation of electrons from the ESR centers. Therefore, the fact that the triplet structure disappears above 40K means that almost all of the ESR centers exhibiting the triplet structure are ionized above 40K. The emitted electrons should work as carriers. However, the carrier density of undoped samples at

40K is less than 1/10 of that at 300K.[5] Therefore, the residual carriers in 3C-SiC at 300K do not result from the ESR centers showing the triplet structure. On the other hand, the broad structure remains even above 40K, although its width becomes broader as the temperature increases.

The ESR spectra of N-doped samples are shown in Fig.3. The intensities are quite high compared to undoped samples. Only broad singlet structure is observed, which is different from the case of undoped samples. The signal intensity decreases and the width becomes broader, as the temperature increases. The appearance of this singlet structure is similar to that of undoped samples. The singlet structure appears on intentional N doping, and should be related to nitrogen atoms. If these ESR centers are located at N atoms, triplet hyperfine structure might be observed. However, there is the possibility of showing broadening on account of the interaction between the ESR centers and conduction electrons, which makes triplet structure unobserved. The broadening during the temperature increase supports this assumption.

In Fig.4, the relation between carrier density at 300K and relative spin density at 4.2K for undoped and N-doped samples is shown. Assuming that the carriers are frozen out at 4.2K, the spin density at 4.2K should correspond to the carrier density at 300K. For N-doped samples, relative spin density is proportional to carrier density, which means that the observed ESR centers correspond to the residual carriers at 300K. However, the spin densities of undoped samples at 4.2K are estimated to be much lower than the extrapolated amounts of ESR centers for the corresponding carrier densities at 300K. These experimental results suggest that the carrier density of undoped 3C-SiC cannot be explained solely by the spins observed at 4.2K.

Considering the results of PL and ESR measurements together, it is indicated at least that the main origin of the residual carriers in undoped 3C-SiC is not the usual nitrogen donor, which is consistent with the results obtained from the study of the electrical properties of N-doped 3C-SiC epilayers.[4] The origins of

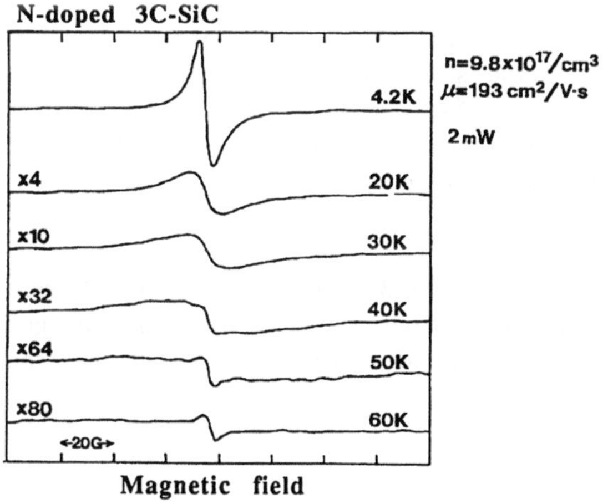

Figure 3. ESR spectra of N-doped 3C-SiC ($n=9.8\times10^{17}/cm^3$) and their temperature dependence.

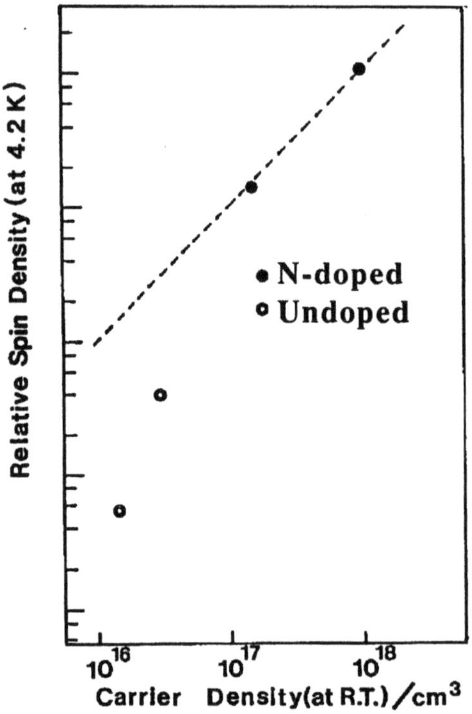

Figure 4. Relation between the carrier density at 300K and relative spin density at 4.2K for undoped and N-doped 3C-SiC epilayers. The spin density is estimated by integrating the measured ESR signals twice including all the components. To suppress the saturation effect, the microwave power for the measurements was around 0.5mW. For N-doped samples, the slope of data points is unity, which means the spin density is proportional to the carrier density.

the residual carriers, which may be observed as the PL emissions, are not clear. We speculate that they may be some structural defects. From the fact that the intensities of these peaks are weak for the samples with low carrier density and high mobility, it is indicated that such defects are more reduced for the undoped 3C-SiC epilayers with low carrier density.

4. Conclusion

PL and ESR spectra of undoped and N-doped 3C-SiC grown by CVD were measured. It was found that undoped 3C-SiC grown by CVD contains nitrogen impurities, however, the main origin of the residual carriers at 300K is not the usual nitrogen donors.

Acknowledgments

We wish to thank S. Ochiai, K. Tanaka and Y. Suzuki for their help in the PL measurements. We are also grateful to S. Maekawa for his continuing encouragement.

References

1) S.Nishino, *et al.*: Appl.Phys.Lett. **42** (1983) 460.
2) W.J.Choyke and L.Patrick: Phys.Rev. **127** (1962) 1868.
3) W.J.Choyke, *et al.*: Phys.Rev. **133** (1064) A1163.
4) M.Yamanaka, *et al.*: Jpn.J.Appl.Phys. **26** (1987) L533.
5) M.Yamanaka, *et al.*: J.Appl.Phys. **61** (1987) 599.
6) H.Okumura, *et al.*: Jpn.J.Appl.Phys. **27** (1988) L116.
7) J.A.Freitas,Jr., *et al.*: J.Appl.Phys. **61** (1987) 2011.
8) H.H.Woodbury and G.W.Ludwig: Phys.Rev. **124** (1961) 1083.
9) Yu.M.Altaiskii, *et al.*: Sov.Phys.-Solid State **12** (1971) 2453.
10) W.E.Carlos, *et al.*: Proc. MRS Symp. on Novel Refractory Semic. **97** (1987) 253.

Phase Matched Second Harmonic Conversion in ß-SiC

K.H. Jackson, E.W. Jones, and G.L. Harris

Materials Science Research Center of Excellence,
Department of Electrical Engineering, Howard University,
Washington, DC 20059, USA

In this article, we describe 2nd harmonic conversion in ß-SiC at a fundamental wavelength of 1.06µm. ß-SiC is a cubic semiconductor crystal with the zinc-blende structure. Ideally such a crystal should be isotropic, i.e. possess no intrinsic birefringence. This lack of birefringence would normally limit the usefulness of ß-SiC for 2nd harmonic conversion, and other applications in nonlinear optics. We have discovered epitaxial layers of ß-SiC formed by vapor phase growth are indeed birefringent, and thus phase matchable by the standard technique of angle tuning. We attribute this to stress induced birefringence (SIB) frozen into the layer during epitaxial growth. Given the 20% lattice mismatch between the Si <100> substrate and the growing ß-SiC layer it is easy to understand why there is so much stress in the layer. Phase matched films of ß-SiC have respectable conversion efficiencies even in the presence of absorption at 532nm. This high conversion efficiency can be understood on the basis of a simple model of the polarizability of the Si-C bond. When compared to crystals of potassium dihydrogen phosphate (KDP) we have determined the 2nd order nonlinear susceptibility of ß-SiC to be $\chi^{(2)}_{ijk} = 1.05 \pm 0.3 \times 10^{-9}$esu.

Introduction:

In recent years considerable research has been focused on the study of ß-SiC for electronic applications.[1] ß-SiC has displayed considerable potential for use in electronic instrumentation which must operate at elevated temperatures or in environments where there are high levels of ionizing radiation. To date, optical studies of SiC have been used to elucidate the band structure of the semiconductor. There has also been a considerable effort to develop blue light emitting diodes, phototransistors, and UV detectors using ß-SiC.[2] This article describes the first observation of phase matched 2nd harmonic generation in an epitaxial film of ß-SiC.

ß-SiC is a crystal with the zinc-blende structure, and a lattice constant of 4.3595 Å. It belongs to the 43m symmetry group, and is similar in this respect to GaAs and other III-V semiconductors. Crystals which belong to the zinc-blende group do not possess inversion symmetry. This would imply the 2nd order nonlinear susceptibility is nonzero. From this consideration alone, 2nd harmonic conversion in ß- SiC would not be unusual. Indeed in analogy with GaAs [3] one would expect 2nd order susceptibility to be quite large, however, since ß-SiC belongs to the cubic group of crystals, it is isotropic, i.e. SiC has no intrinsic birefringence. This lack of birefringence would limit the use of SiC for 2nd harmonic gener-

Springer Proceedings in Physics, Vol. 34 **Amorphous and Crystalline Silicon Carbide**
Editors: G.L. Harris and C.Y.-W. Yang

ation, since it is not possible to satisfy the phase matching condition associated with this process.[4-5] Without phase matching, the conversion efficiency for 2nd harmonic generation in GaAs, and other cubic crystals is very small. Although one would expect the 2nd order nonlinear susceptibility of ß-SiC to be nonzero, there is no reason a priori to expect a reasonable conversion efficiency due to the lack of phase matching.

We have discovered it is possible to produce epitaxial films of ß-SiC which are capable of very high conversion efficiencies (comparable to KDP at 1.06μm), and which are phase matchable. Phase matching is possible in ß-SiC due to the large amount of frozen stress in the epitaxial film. This stress causes the film to behave like a uniaxial crystal through the photoelastic effect .[6] The origin of this stress induced birefringence (SIB) is the 20% lattice mismatch between ß-SiC, and the silicon <100> substrate upon which it is grown. In order to accommodate the silicon lattice the atoms in ß-SiC must adjust their equilibrium positions, resulting in a large amount of stress which persists for many microns away from the Si/SiC interface. The SiC films produced are birefringent, and phase matching by angle tuning is now possible. ß-SiC has a large bandgap[7] (Eg = 2.224 eV at room temperature) which allows 2nd harmonic conversion in the visible region, up to its absorption edge at 539nm. Harmonic conversion is possible at somewhat shorter wavelengths because the optical absorption of ß-SiC does not increase rapidly with decreasing wavelength, thus harmonic conversion is possible down to 532nm with respectable efficiencies. Another attractive feature of ß-SiC for harmonic conversion is its high damage threshold, greater than 1 GW/cm^2 for surface damage to a 30 μm film at a incident wavelength of 1.06 μm.

Experimental:

The films used in this work were grown in the Howard University Vapor Phase Epitaxial Growth Reactor. Details of the reactor can be found in the reference section.[8] The films were unintentionally doped n-type, and were grown both on axis, and 2° "off axis" , Si <100> substrates. These off axis substrates have shown substantial advantages over conventional Si substrates in reduction of antiphase boundaries in epitaxial ß-SiC films.[9] The films used in this work were 30μm thick as measured using the sectioning technique on sacrificial substrates.[10] The SiC films were removed from the silicon substrates by dissolving the silicon in a combination of one part HNO_3 to one part HF for 20 minutes. The SiC layer was then washed in deionized water for several minutes. The free-standing films were inspected for good surface morphology and mounted on a glass slide with a 0.5 cm round opening. This provided for easy handling of the 30μm film and clear access for the laser beam. Radiation from a CW mode locked Nd:YAG laser was used as the source of 1.06 μm radiation in the experiment. The output of the laser consisted of 100 psec pulses with a repetition rate of 38 MHz. The pulses have a peak power of 30kW with a total average power of 10 watts. (The experimental apparatus is shown in Figure 1.) The output of the laser was then passed through a variable attenuator, a chopper, a 1.06μm 1/2 wave plate, and focusing lens. The sample was placed in a rotation stage at the focal point of a

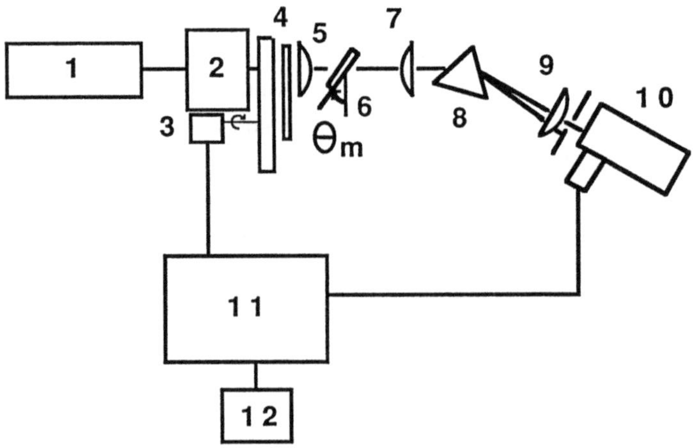

FIG. 1. The experimental arrangement used to observe 2nd harmonic generation in ß-SiC; 1. Nd:YAG laser, 2. beam attenuator, 3. chopper, 4. 1/2 wave plate, 5. lens 1, 6. sample position, 7. lens 2, 8. dispersing prism, 9.lens 3 and iris, 10. monochromator and photomultiplier, 11. phase sensitive detector, 12. oscilloscope. The angle adjusted for phase matching q_m is also shown.

75mm lens. Light was collected from the sample with another 75mm lens and dispersed through a prism with a 60° apex angle. The prism was used to separate the fundamental beam, at 1.06µm from the 2nd harmonic at 532nm. Light from the prism was collected with a 50mm lens and sent through an iris. The iris was adjusted to completely block the fundamental beam at 1.06µm and allow the 532nm light onto the entrance slits of a 1/4 meter monochromator. A photomultiplier tube was placed at the exit slits of the monochromator, and connected to a phase sensitive detector (PSD). The reference frequency of the (PSD) was provided by the chopper. The photomultiplier tube had a GaAs photocathode (RCA C31034) which has a high quantum efficiency 10%, for 532nm radiation and good rejection for any radiation at 1.06µm (QE = .01%). Considerable care was taken throughout the experiment to ensure that no 1.06 µm radiation, either from the direct beam or stray reflections from the various optical surfaces, was incident on the monochromator. This step was very important, because any 1.06µm radiation which was incident on the monochromator could yield a signal in 2nd order and introduce error in the measurement of the 532 nm signal. This is a particular problem since we are attempting to measure a relatively weak 532 nm signal in the presence of a strong 1.06µm beam. To verify that we were measuring the 532 nm beam from our sample of SiC, free of any 1.06µm radiation, the sample was removed from the beam path. Without the sample in place no 532 nm signal was observed. The monochromator was operated at 532 nm with a slit width of 70 µm. This provided a spectral resolution of approximately 5nm. The photomultiplier with the anode directly coupled to the (PSD) was operated at 1000 volts.

Results and Discussion

The average power incident on the sample had a maximum value of 2 watts. This was focused to the diffraction limit, onto the ß-SiC sample, creating a total average power density at the sample of 50MW/cm^2. The 30 µm film showed no signs of surface damage even after being exposed to the laser beam for several days. The maximum 532 nm signal was observed when the fundamental beam was at a 60° angle w.r.t. the optical axis of the SiC film. The signal intensity at 60° was four times larger than the signal at normal incidence. The input beam was polarized at a angle of 70° w.r.t. the vertical. Figure (**1**) shows the experimental geometry. A plot of the variation of intensity vs. angle is shown in Figure (**2**).

The sample was rotated to the observed phase match angle and the variation of the relative intensity at 532nm vs. the relative intensity at 1.06 µm was made; the results of these measurements are shown in Figure (**3**). There is a quadratic relationship between the fundamental power at 1.06 and the 532 nm power. The data shown represent the average between four different measurements, and both the 532nm and 1.06µm intensities have been normalized w.r.t. their maximum values. The quadratic increase in 532nm output with 1.06µm intensity is a well-known signature of 2nd harmonic generation.

To measure the damage threshold of the layer the full power of the laser (10 watts) was allowed to strike the sample. The 10 watts of laser radiation were focused on the sample with a total power density of 1 GW/cm^2. After several minutes there was no surface damage observable on the sample. To further insure we were in fact observing 2nd harmonic generation, the 532nm signal was measured when the laser was allowed to operate in continuous wave (CW) mode. In CW mode the 532nm signal was not present. This confirms

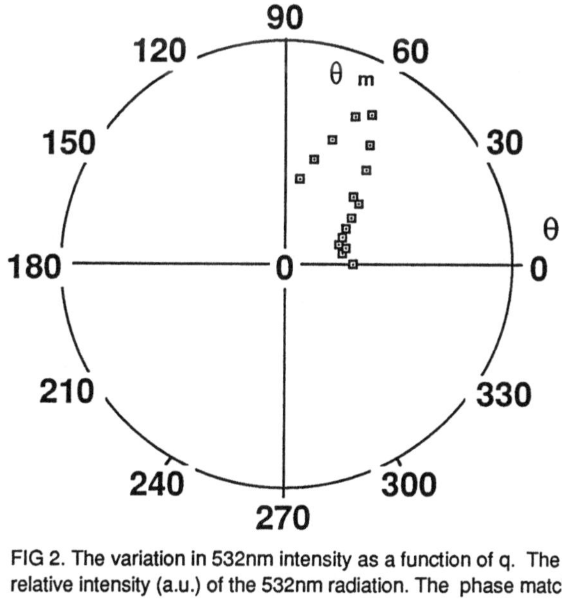

FIG 2. The variation in 532nm intensity as a function of q. The radius measured to the plotted points is the relative intensity (a.u.) of the 532nm radiation. The phase match angle for ß-SiC was qm = 60° .

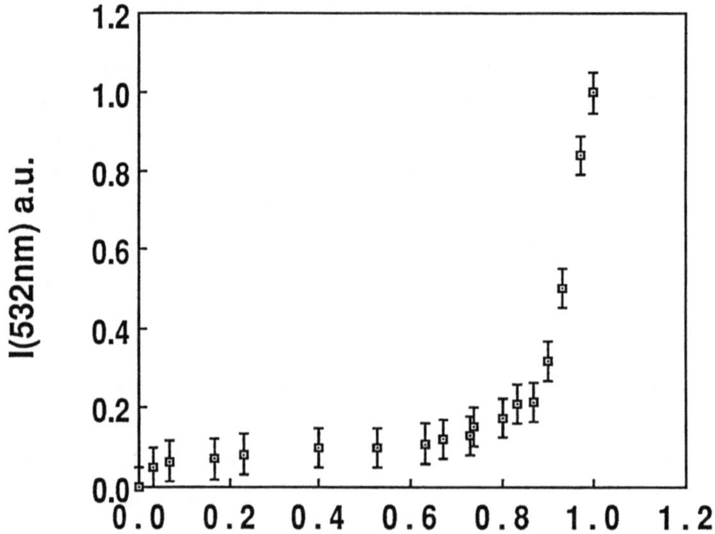

FIG 3. The variation of 2nd harmonic intensity (532nm) with fundamental intensity (1.06μm). The 532nm output displays a quadratic dependence on 1.06 μm intensity, a classic signature of 2nd harmonic generation.

the 532nm signal was dependent on the peak, not the average power of the laser, as one would expect in 2nd harmonic generation.

In order to get a comparison of the efficiency of 2nd harmonic generation in ß - SiC we compared the 2nd harmonic efficiency with KDP. Table 1 shows the result of this comparison. In both cases the crystals were measured under phase matched conditions. We have also made a correction for focused gaussian beams [11], hence the efficiency is linear in the crystal length, not quadratic as in the case of plane waves. In addition, the surfaces of the SiC sample were not optically polished, as was the case for the KDP crystal. There was a significant amount of 532nm light scattered out of the field of view of the monochromator. This scattering was severe enough to allow the 532nm radiation to be observed visually, perpendicular to the optical axis. Thus, we feel our estimate of the 2nd harmonic conversion efficiency is a very conservative one. Using the comparison with KDP [12] we estimate the 2nd order nonlinear susceptibility of ß-SiC to have a magnitude of $\chi^{(2)}_{ijk}$ = $1.05 \pm 0.3 \times 10^{-9}$ esu. The large conversion efficiency of ß-SiC is even more striking when one considers that our measurement did not take into consideration the absorption of 532nm radiation within the film itself. This high conversion efficiency can be understood by carefully studying the ionic character of the covalent bonds of tetrahedrally coordinated semiconductors. Tang and Phillips, on the basis of simple molecular models, have related the magnitude of the 2nd order nonlinear susceptibility to the degree of ionic character of the A-B bond, and the degree of charge transfer between atoms A and B.[13-14] The relationship may be expressed as [15]

$$\chi^{(2)}_{ijk} = K[\ f_i\,(1-f_i)^5\ ]^{1/2}\ . \qquad (1)$$

Table I. 2nd harmonic conversion efficiency of ß-SiC compared to KH_2PO_4 (KDP) at a fundamental wavelength of 1.06μm.

Material	Length	Observed[a] Efficiency	Corrected[b] Efficiency	$\chi^{(2)}_{ijk}$[c] (x 10^{-9} esu)
KDP				
	300μm	1.0	1.0	2.34
ß-SiC				
	30μm	0.045	0.45	1.05

(a) The crystals were rotated to their phase matched positions, and the relative intensity of radiation at 532nm was compared to that observed for KDP at the same 1.06μm intensity, i.e., $I_{SiC}(532nm)/I_{KDP}(532nm)$ = Observed Efficiency.

(b) The correction for crystal length was made assuming Gaussian focusing ref. (11), under these conditions the efficiency increases linearly with crystal length.

(c) The value for the 2nd order susceptibility, $\chi^{(2)}_{ijk}$, was taken from ref. (12).

K is a constant of proportionality, and f_i is defined as the fraction of ionic or hetropolar character in the bond compared to the fraction f_h of covalent or homopolar character. Clearly these fractions must satisfy the relation

$$f_i + f_h = 1.00 \quad . \qquad (2)$$

The maximum value for $\chi^{(2)}_{ijk}$ is observed when f_i is equal to 1/6 =.1667. Table 2 compares the magnitude of f_i for SiC with some other semiconductors. Most remarkably, ß-SiC is the only zinc-blende crystal whose single bond ionicity is close to the value which would maximize the 2nd order susceptibility $\chi^{(2)}_{ijk}$. Calculations of the nonlinear susceptibility in semiconductors based upon the single bond ionicity have been very successful.[16-17] These same models would predict that ß-SiC should have a very large conversion efficiency, *perhaps the largest of any zinc-blende type crystal* ; our observations seem to indicate this is indeed the case.

Phase Matching

Consider a thin film which has forces applied uniformly over the thickness at its edges. Under these conditions we assume that the principal stress s_{xx} and s_{yy} in the plane of

Table II. Magnitude of single bond ionicity f_i (dimensionless) , and covalency f_h of some tetrahedrally coordinated semiconductors[(a)].

Semiconductor Crystal	f_i	f_h
Diamond	0.0	1.00
Si	0.0	1.00
AlN	0.449	0.551
GaAs	0.310	0.690
GaP	0.327	0.673
ß-SiC	0.177	0.823

(a) Ref. (15) p. 42

the film are constant throughout the thickness. The relative phase difference introduced between orthogonally polarized components is [18]

$$\delta = \pi d/\lambda n_o^3(q_{11}-q_{12})\,(\sigma_{xx} - \sigma_{yy}) \qquad (3)$$

where n_o is the refractive index of the material, q_{11}-q_{12} the stress optical coefficients, and d is the thickness of the layer. Compare equation (3) to the phase difference introduced by a uniaxial crystal, [19]

$$\delta = (n_e-n_o)\ 2\pi d/\lambda\ . \qquad (4)$$

Thus, the stress frozen into the ß-SiC layer provides enough birefringence to allow an otherwise isotropic material to behave like a uniaxial crystal. We have measured the optical sign of the stressed film and found it to be a positive uniaxial .[20]

The data shown in figure (2) clearly show a dependence on the phase matching angle. The phase matching angle is not as peaked as is predicted by theory. This is due to absorption within the film at 532nm. Some absorption at this wavelength is to be expected since the energy of the 532nm radiation (hv=2.33 eV) is greater than the bandgap E_g, however, ß-SiC is an indirect gap material and the fundamental absorption increases slowly with increasing photon energy (decreasing wavelength). In the presence of absorption at 532nm, the intensity of the 2nd harmonic output is given by [21]

$$I(2\omega)= [\ 2\pi\ (\ 2\omega\ )^2/c\sqrt{\varepsilon}\ ]\ \{\ \chi^{(2)}{}_{ijk}\ \}^2\ I^2(\omega)/[(\Delta k)^2 + \beta^2] \qquad (5)$$

where ß is the attenuation coefficient at 532nm. The phase matching condition is met when $\Delta k=2k_1-k_2=0$. In the absence of absorption, $I(2\omega) \propto \{\sin(\Delta kz/2)/\Delta kz/2\}^2$,and is strongly peaked when $\Delta k = 0$. The effect of absorption is to broaden the peak half width to $(1/\beta z)^2$, and reduce its maximum value. With absorption the effective interaction length is reduced to 1/ß.

To summarize, we have observed efficient phase matched 2nd harmonic generation in ß-SiC at 1.06µm. We have attributed the phase matching behavior, in what would otherwise be an isotropic crystal, to the stress induced birefringence (SIB) in the SiC epilayer. This (SIB) is due to the lattice mismatch between the ß-SiC and Si substrate, which is frozen in the film during the growth process. The as-grown epitaxial layer behaves like a positive uniaxial crystal which can be phased matched in the standard manner. The observed efficiency can be understood on the basis of the single bond ionicity of ß-SiC. Indeed, of all the crystals with the zinc-blende structure, SiC has almost the optimal value of ionicity to maximize the 2nd harmonic output. It is our view that the significance of ß-SiC as a material for special purpose electronic applications will be superseded by its use in nonlinear optics in the near future. This work was supported by NSF RII-8413805, NSF R11-8714767, NASA Lewis NAG-3-431, and a grant from Standard Oil of Ohio.

References

1. S. Nishino, J. A. Powell, H. A. Will, Appl. Phys. Lett. 42, 460 (1983)
2. L. Hoffmann, G. Ziegler, D. Theis, and C. Weyrich, J. Appl. Phys. 53, 6962 (1982)
3. The published value of the 2nd order nonlinear susceptibility for GaAs is $\chi^{(2)}_{ijk}$= 377± 38 x (3/4π x 10^{-8} esu). This value may be found in R. J. Pressley,ed., *Handbook of Lasers* (Chemical Rubber Co., Cleveland, Ohio, 1971) p.497.
4. F. Zernike, *Methods of Experimental Physics, Quantum Electronics*, Vol 15B, ed. by C.L. Tang, (Academic Press, New York 1979), p. 152.
5. J. Wynne, and N. Bloembergen, Phys. Rev. Lett. 16, 613 (1969).
6. M. Born and E. Wolf, *Principles of Optics*, 6th ed. (Pergamon Press, Oxford 1980), p. 703.
7. M. Neuberger, *Handbook of Electronic Materials*, Vol 5, Group IV Semiconductor Materials. (IFI/Plenum, New York, 1971), p. 61-62.
The value of the energy gap at 300K was calculated using the 4.2K value given in (7a) and using a formula for the change of the energy gap with temperature given in reference (7b).
(7a). W. J. Choyke, D.R. Hamilton,L. Patrick, Phys. Rev. 133, A1163 (1964)
(7b). M.L. Belle et al., Sov. Phys.-Semicond. 1, 315 (1967) Eg(T=300K)= 2.39eV-(2.2×10^{-4} eV/K) (300K) = 2.224 eV
8. G. L. Harris, K. H. Jackson, G. J. Felton, K.R. Osborne, K. Fekade, and M. G. Spencer, Mater. Lett. 4, 77 (1986).
9. K. Shibahara, T. Saito, S. Nishino, and H. Matsunami, IEEE Electr. Dev. Lett., EDL-7, (1986)
10. W. W. Hoff, W. Shockley, Bull. Am. Phys. Soc. 11, 382 (1956).
11. A. Yariv, *Optical Electronics* , (Holt, Rinehart and Winston, New York 1985), p. 243.

12.The published value of the 2nd order nonlinear susceptibility for KDP is $\chi^{(2)}_{ijk}$= 0.98± 0.04 x (3/4π x 10^{-8} esu) see reference (3) .
13. C. L. Tang, IEEE J. Quant. Electr. QE-9, 755, (1973)
14. J. C. Phillips, J. A. VanVachten, Phys. Rev. 183, 709 (1969)
15. J. C. Phillips, *Bonds and Bands in Semiconductors*, (Academic Press, New York 1973), 182.
16. C. L. Tang, C. Flytzanis, Phys. Rev.B4, 2520 (1971)
17. D. A. Kleinman, Phys. Rev. B2, 3139 (1970).
18. Reference (6) p.705 .
19. Ibid p. 703.
20. K. H. Jackson et al. " The Measurement of Interfacial Stress Using the Photoelastic Effect", talk given at the Materials Research Conference, June 24, 1987 Santa Barbara, CA.
21. Y. R. Shen, *Principles of Nonlinear Optics* , (Wiley-Interscience, New York 1984), p. 76.

Gamma-Ray Irradiation Effects on 3C-SiC Devices

I. Nashiyama, M. Shinohara, T. Matsumoto, K. Endo, E. Sakuma, S. Misawa, H. Okumura, and S. Yoshida

Electrotechnical Laboratory,
1-1-4 Umezono, Tsukuba, Ibaraki 305, Japan

Radiation hardness of 3C-SiC MOSFET's, MESFET's and PN-junction diodes were studied by ^{60}Co gamma-ray irradiations up to $1x10^6$ rads. The MOSFET's and the MESFET's were found to be tolerant against gamma-ray irradiations up to $1x10^6$ and $1x10^7$ rads, respectively. Significant recoveries were observed for both transistors after 47 hours' annealing at room temperature. Irradiation effects on device characteristics depend on the biasing condition during irradiation. The PN-junction diodes showed no degradation up to $1x10^7$ rads.

1 Introduction

Silicon carbide is an excellent candidate material for high temperature semiconductor devices, because of its wide band gap and high electron mobility. Moreover, SiC electronic devices are expected to be tolerant against ionizing radiations, since the rate of the electron-hole pair generation induced by irradiation in SiC is much lower than that in silicon because of its wider band gap and lower atomic number of the element atoms. However, very little is known about radiation effects in semiconducting SiC crystals or in SiC devices. Babcock[1)] reviewed several studies of irradiation damage in silicon carbide, and discussed degradation in charactreristics of SiC nuclear radiation detectors irradiated with high energy charged particles and neutrons. He pointed out that the experimental values were scattered over more than an order of magnitude, and that better material control was a prerequisite to the understanding of radiation-induced defects in silicon carbide.

Recently, we have successfully developed 3C-SiC single crystal layers[2,3)] and fabricated several 3C-SiC devices such as MOSFET's[4)], MESFET's[5)], and PN-junction diodes[6,7)]. In the present study, radiation hardness of these electronic devices was examined by applying ^{60}Co gamma-ray irradiation up to $1x10^7$ rads.

Springer Proceedings in Physics, Vol. 34 Amorphous and Crystalline Silicon Carbide
Editors: G.L. Harris and C.Y.-W. Yang

2 Experiment

Details of the 3C-SiC device fabrication methods and their high temperature characteristics are described elsewhere.[2-7]. The outline of the device fabrication procedure is as follows. A boron-doped semi-insulating 3C-SiC layer of about 5 µm thickness was epitaxially grown on a p-type Si(100) substrate, and a thin undoped 3C-SiC epilayer of about 650 nm was successively grown on the boron-doped epilayer. The boron-doped epilayer has a resistivity more than 100 Ωcm at room temperature, which electrically isolates the undoped SiC layer from the silicon substrate. The undoped 3C-SiC epilayer shows n-type conduction, having a resistivity of 0.3 Ωcm, a carrier concentration of 3×10^{16} cm^{-3}, and a Hall mobility of 700 cm^2/Vs at room temperature typically. In the case of the MOSFET's, the surface of the undoped n-type layer was thermally oxidized at 1050 C for 4 hours in oxygen. The thickness of the oxide layers is about 60nm. Both types of transistors were fabricated using this 3C-SiC epitaxial double layer. Since the same electrode pattern was used for both MOSFET's and MESFET's, they have exactly the same top view (Fig. 1(a)). Schematics of the cross-sectional views of the MOSFET's and the MESFET's are shown in Figs. 1(b) and (c), respectively. Both transistors operate in an n-channel depletion mode. The PN-junction diode has a mesa structure where an undoped 3C-SiC layer and an aluminum doped p-type 3C-SiC layer are successively grown on a Si(100) substrate to form a SiC PN junction.

The 3C-SiC devices were subjected to ^{60}Co gamma-ray irradiation up to 1×10^7 rads, and subsequent room temperature annealing. The irradiation test procedure was based on MIL STD 883B method 1019.1. The specimen devices were covered with 2.5 mm thick aluminum in

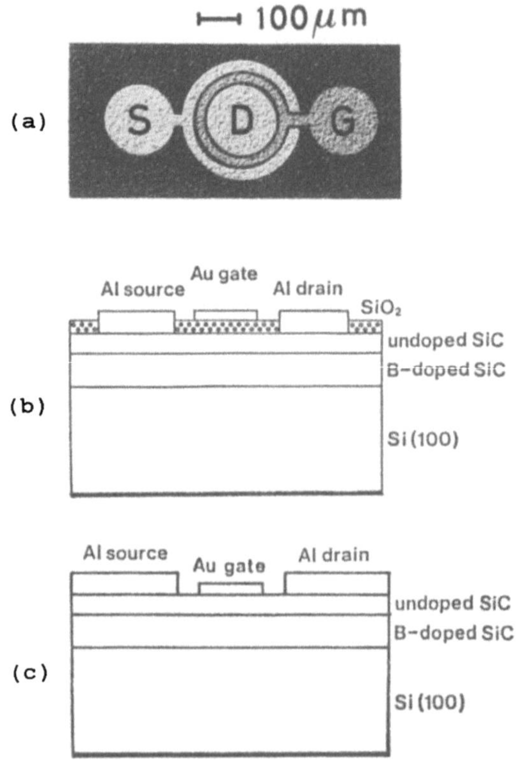

Fig.1. (a)Top view of 3C-SiC MOS and MES FET. S:Al-source,D:Al-drain, G:Au-gate. Schematics of (b)MOSFET structure and (c)MESFET structure.

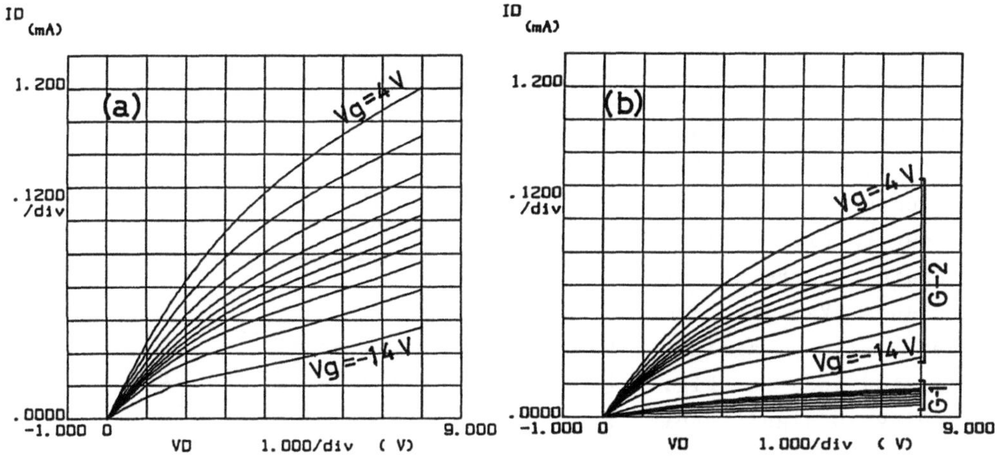

Fig.2. Drain current-voltage characteristics(Vg=4V to -14V step=-2V) of 3C-SiC MOSFET,(a)before and (b)after gamma-ray irradiation with all electrodes floated. G-1:10 Mrads, G-2:annealed at 25°C for 47 hours.

order to generate a consistent secondary electron fluence on the devices. Electrical parameters were measured immediately before and immediately after irradiation. Biasing conditions during irradiation were as follows. 1)positive, 2)negative, 3)zero bias was applied to the gate with the other electrodes being grounded, 4)only the gate was floated with the other electrodes being grounded, 5)all the electrodes were floated. All the measurements and irradiations were performed at room temperature.

Results and discussion

Figure 2(a) shows typical drain current-(Id)-drain voltage(Vd) characteristics of the 3C-SiC MOSFET before irradiation, where gate voltage(Vg) was applied from 4 to -14 V with -2 V step. At Vd=4V, the transconductance(Gm) was 1.8 mS/mm, and the threshold voltage(Vt) was -18.6V, obtained from a relation $Id=\beta(Vg-Vt)^2$.

By gamma - ray irradiation, the drain

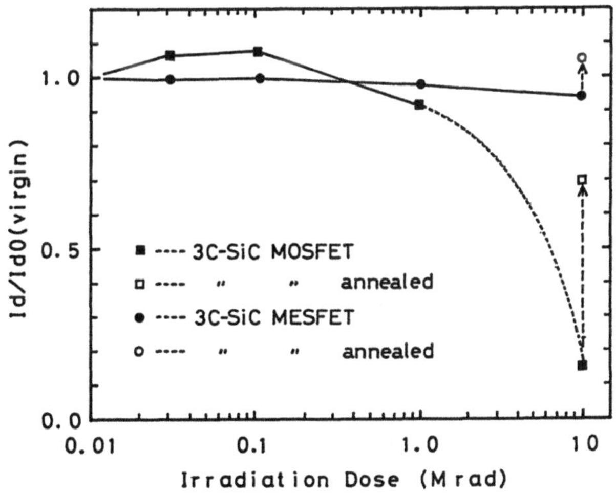

Fig.3. Normalized drain current plotted against irradiation dose for MOSFET and MESFET. All the electrodes were floated during irradiation.

current decreased with increasing irradiation dose. The typical Id-Vd characteristics after 1×10^7 rads irradiation and successive annealing are shown in Fig. 2(b). The drain currents at Vd=8V and Vg=-14V are plotted against irradiation dose in Fig. 3, showing no significant reduction in Id at a gamma-ray dose up to 1×10^6 rads. For higher doses, Id decreases rapidly to be an order of magnitude lower than its initial value at 1×10^7 rads, resulting in marked reductions of both Gm and the channel conductance (Gd). It should be noticed, however, that this marked degradation recovered to a very great extent by 47 hours' annealing at room temperature as shown in Figs. 2(b) and 3. Figure 4 shows the irradiation dose dependences of Vt and Gm under the various biasing

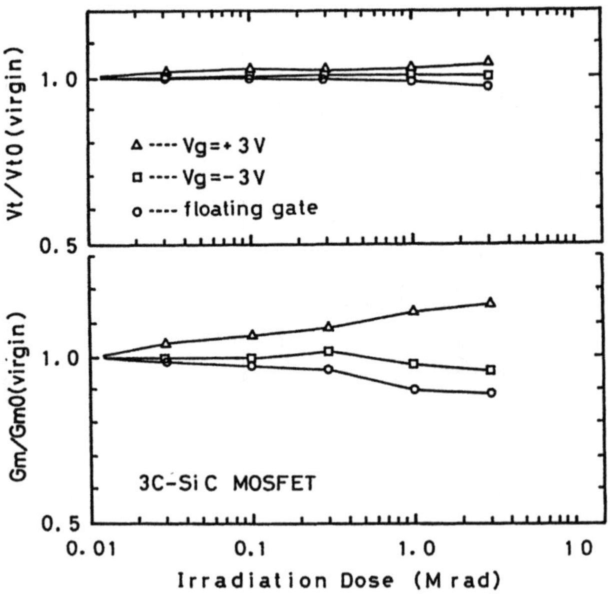

Fig.4. Effect of biasing during irradiation on Vt and Gm of 3C-SiC MOSFET. Vt and Gm are normalized to their initial values.

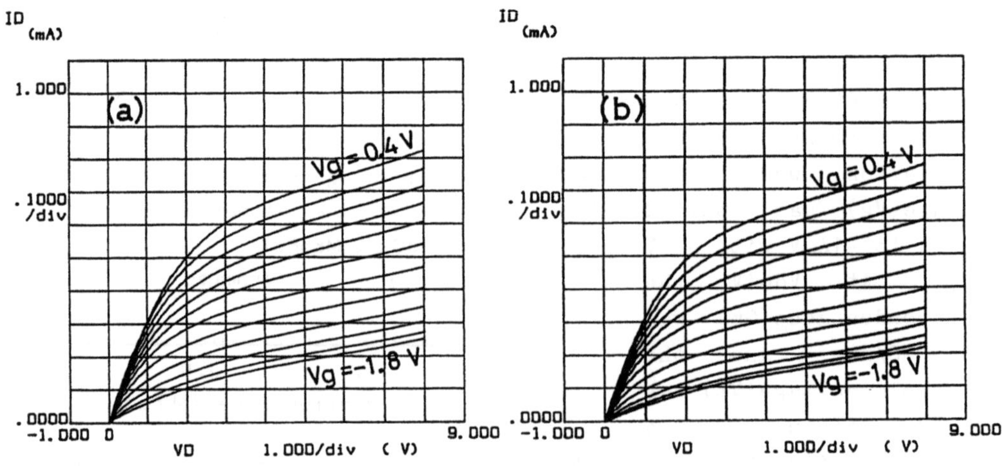

Fig.5. Id-Vd curves(Vg=-0.2 to -1.8V, step=-0.2V) of 3C-SiC MESFET, (a)before and (b)after 10Mrads irradiation with all electrodes floated

conditions during irradiation. In the case of the positive bias irradiation, Gm increases with irradiation dose, whereas it decreases in the case of the floating gate irradiation. Then, Gm became approximately 115% and 90% of its initial value respectively at a dose level of $1x10^7$ rads. The threshold voltage does not vary so much with gamma-ray irradiation. The biasing effects will be discussed later in terms of the damage mechanism.

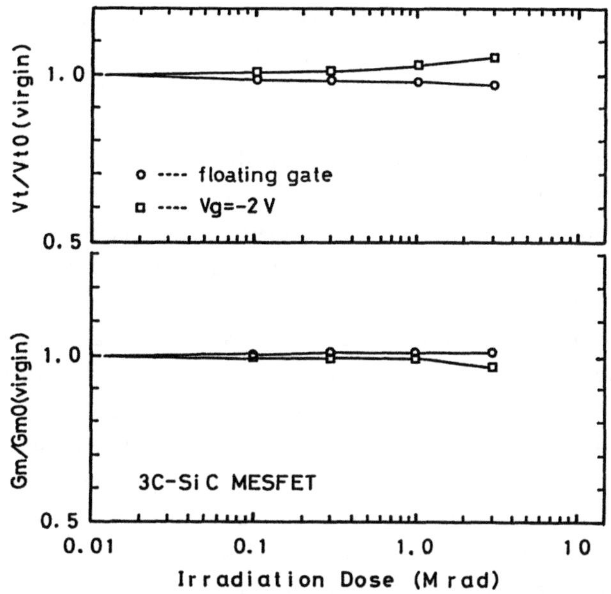

Fig.6. Effect of biasing during irradiation on Vt and Gm of 3C-SiC MESFET. Vt and Gm are normalized to their initial values.

Similar irradiation tests were performed for 3C-SiC MESFET's, where Vg was applied from 0.4V to -1.8V with a step of -0.2V. Results are shown in Figs. 5(a) and 6(b), where only slight decrease in Id was observed even after $1x10^7$ rads irradiation. The drain currents at Vd=8V and Vg=0.4V are plotted against irradiation dose in Figure 3, indicating that the MESFET exibits much stronger tolerance against gamma-ray irradiation than the MOSFET. Figure 3 show a significant recovery (more than 100%) of Id also for the MESFET by 47 hours' annealing at room temperature after $1x10^7$ rads irradiation, and Id exceeds its initial value. The biasing effects for the MESFET are smaller than those for the MOSFET as shown in Fig. 6. PN-junction diodes showed no degradation up to $1x10^7$ rads.

Ionizing radiation effects on the 3C-SiC MOSFET might be understood by the combination of two basic mechanisms, 1) positive charge buildup near the SiC-SiO_2 interface, and 2) effects of the negative charges trapped at the interface between the undoped SiC layer and the semi-insulating SiC layer. The first mechanism is similar to the radiation damage mechanism in silicon MOSFET, which takes place mostly when the gate is positively biased during irradiation. Irradiation induced positive charges in the oxide drift toward the SiO_2-SiC interface and caused positive charge buildup near the interface. These positive charges decrease the gate depletion layer depth and increase Id and Gm as shown in Fig. 4. This mechanism may be dominant at lower dose levels. The second mechanism might be peculiar to the present device structure. The ionizing radiation introduces a large amount of electron-hole pairs in the thick semi-insulating SiC layer. The induced electrons may move to the n-SiC/i-SiC interface, being trapped there by the interface states. This varies the depth of the depletion layer at the n-SiC/i-SiC interface, causing significant decrease in drain current. The second mechanism sometimes surpasses the first mechanism at high dose levels above $1x10^6$ rads. These trapped charges seem to be unstable and are easily swept away at room temperature, which results in the recovery of Id as shown in Fig.3. Ionizing radiation effects in semiconductor devices usually depend strongly on both device structures and fabrication methods. Therefore, the present results for the 3C-SiC devices cannot be directly applied to other SiC devices. However, the possibility of new radiation-hard devices was indicated, which can be used also in an elevated temperature environment.

Conclusion

The depletion mode 3C-SiC MOSFET's and MESFET's were found to be tolerant against ^{60}Co gamma-ray irradiation up to $1x10^6$ rads and $1x10^7$ rads, respectively. The irradiation effects depend on the biasing conditions during irradiation. Degradations of the MOSFET's and MESFET's recover significantly by room temperature annealing. PN-junction diodes are insensitive to irradiation up to $1x10^7$ rads.

References

1) R. Babcock: IEEE Trans. on Nuclear Sci. NS-12, No.6 (1965)43.
2) K. Sasaki, E. Sakuma, S. Misawa, S. Yoshida, and S. Gonda: Appl. Phys. Lett. 40(1984)72.
3) M.Yamanaka, H.Daimon, E.Sakuma, S.Misawa, and S.Yoshida: J. Appl. Phys. 61(1987)599.
4) S.Yoshida, K.Endo, E.Sakuma, S.Misawa, H.Okumura, H.Daimon, and E.Muneyama, and M.Yamanaka: Proc. of the spring meeting of MRS 87' (to be published).
5) H.Daimon, M.Yamanaka, M.Shinohara, E.Sakuma, S.Misawa, K.Endo, and S.Yoshida: Appl. Phys. Lett. (to be published).
6) S.Yoshida, K.Sasaki, E.Sakuma, S.Misawa, and S.Gonda: Appl. Phys. Lett. 46(1985)766.
7) H.Daimon, M.Yamanaka, E.Sakuma,S.Misawa and S.Yoshida: Japan. J. Appl. Phys. 25(1986)L592.

The Measurement of Stress in Silicon Carbide Using the Photoelastic Effect

K.H. Jackson, P. Zhou, E. Jones, G.L. Harris, and M.G. Spencer

Materials Science Research Center of Excellence,
Department of Electrical Engineering, Howard University,
Washington, DC 20059, USA

The authors have developed a novel technique for the measurement of stress in thin epitaxial layers of semiconductor material systems. The experimental set-up can best be described as a high resolution, photoelastic stress analysis system. In this paper, results are presented for measurements of SiC grown on Si (100), and 2° Si (100) off-axis in the (011) direction, as well as GaAs and AlGaAs systems.

Introduction

As far back as 1852, Maxwell [1] described a theory whereby the amount of polarized light which gets through a material is proportional to the amount of stress in the sample. An important technique for the measurement of stress-induced birefringence in a variety of technologically interesting thin film and bulk semiconductor materials was designed by researchers at the Howard University's Materials Science Research Center of Excellence. The method utilized can best be described as high resolution photoelastic stress analysis [2]. The technique is very simple. An apparatus is set up to include a laser, two crossed polarizers and several optical devices. A semiconductor sample which needs to be examined for evidence of stress is placed between the two crossed polarizers. The laser provides a source of radiation to which the semiconductor sample is transparent. The incident laser light is extinguished by the crossed polarizers when no stress is detected in the sample. If stress is present in the sample, a small amount of birefringence is induced by the stress, which rotates the plane of the polarized light. This rotation allows light to pass through the second polarizer and reach a detector. The amount of light which can be detected is proportional to the amount of stress in the material.

Chemical Vapor Deposition (CVD) and Molecular Beam Epitaxy (MBE) technology has made it possible to grow very thin layers of highly strained material. An epilayer may be only 100Å thick and may contain large amounts of stress due to the lattice mismatch between layers. This phenomenon makes the materials suitable candidates for the type of measurement obtained by the technique described here. To use photoelastic stress analysis for the study of extremely thin epitaxial layers or bulk semiconductor wafers, the sensitivity must be increased considerably. What is unique about this new apparatus, when compared to commercially available photoelastic systems, is that the overall experimental sensitivity is greatly increased due to the inclusion of phase sensitive detection and a noise reduction scheme which employs a notch filter.

The apparatus described above is capable of measuring strains as small as 10^{-7}. The foundation of the system is based on the idea that the amount of stress-induced birefringence in a sample is converted into a sinusoidal signal at a known frequency. This signal is measured using a phase sensitive photodetector. This technique is useful because it is provides highly quantitative results, yet is very simple to use for a variety of materials.

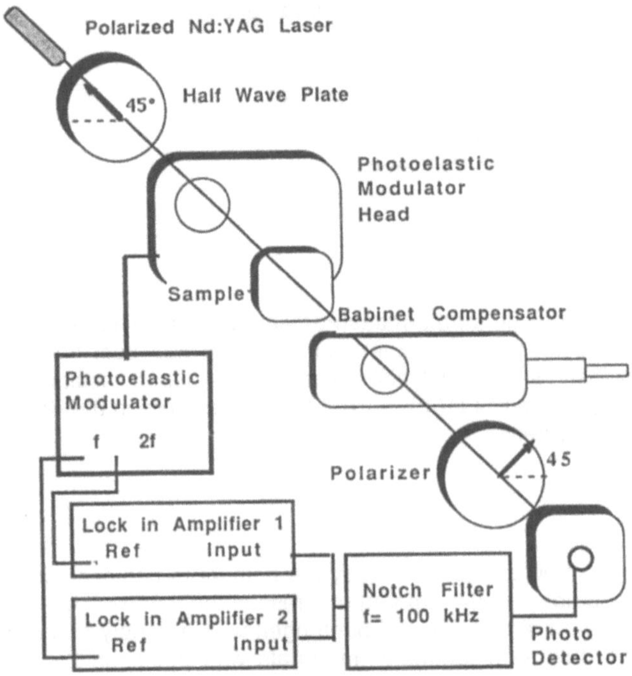

Figure 1: Optical arrangement for photoelastic stress measurement

Experimental Apparatus and Results

The experimental set-up shown in Figure 1 was designed to increase the sensitivity of photoelastic analysis so that the amount of stress in thin epitaxial layers or bulk wafers could be detected and measured. The light source is a linearly polarized Nd:YAG laser operating at 1.06 µm. This laser provides subbandgap radiation for the study of SiC and other materials (provided that $E_g > E_\lambda$). The half-wave plate allows for the convenient rotation of the plane of polarization of the Nd:YAG laser. The photoelastic modulator is used to vary the polarization state of the incoming radiation before it reaches the sample. The compensator is used to nullify any stray birefringence caused by the optics and to provide quarter-wave retardation for calibration purposes. The output signal is then sent through an analyzer to the photodetector. The signal from the photodetector is analyzed by two lock-in amplifiers. Consider the retardance and output signal intensity, Fig. 2a, with the sample removed from its position in Fig. 1, the output intensity is modulated at 100kHz. When the compensator is adjusted for quarter-wave retardation the output intensity is modulated at 50 kHz, Fig. 2. The component of the intensity observed at 50 kHz is proportional to the amount of stress induced birefringence present in the sample. The maximum amount of retardation is quarter-wave ($\pi/2$), therefore the instrument can be calibrated by inserting a fixed 1/4 wave plate in the sample position, or by adjusting a compensator for 1/4-wave retardation[3].

The photoelastic stress measurement technique was used to study thin films of silicon carbide grown on silicon. The silicon carbide (ß-type) was grown in a horizontal reactor employing a water cooled quartz jacket. Under standard conditions the growth takes place at 200 torr using a two-step process involving a carbonized transition layer followed by chemical vapor deposition of SiC. More details on the growth system and conditions have been described elsewhere [4].

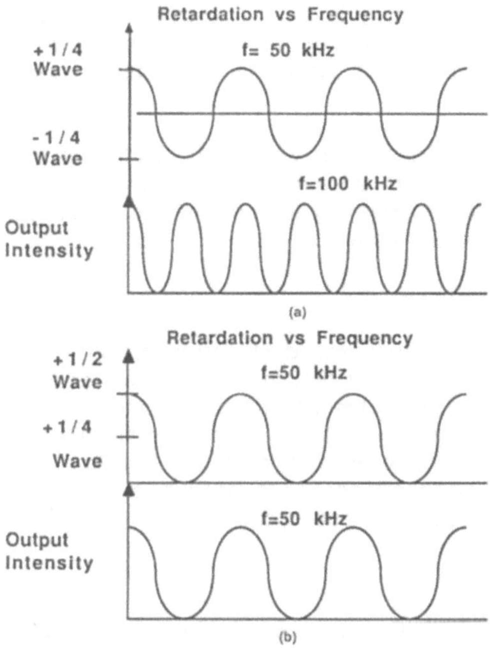

Figure 2 : (a) Plot of retardation, and output intensity versus frequency without sample present. (b) Plot of retardation, and output intensity versus frequency with quarter-wave plate in sample position. Note that the output is now modulated at 50 kHz.

The two inch crystals of β-SiC were grown on Si(100) and Si(100) 2° off axis towards the (011) direction. The ratio of Si/C in the gas phase was held at 0.7. The wafers were cut into 0.25 inch squares and thinned using an $HF:HNO_3$ etch. The resulting SiC films ranged in thickness from 10 µm to 40 µm and were mounted on a glass slide with a hole in the middle to allow the transmission of 1.06 µm radiation without any attenuation.

The results of SiC measurements are shown in Table 1. GaAs and GaAlAs/GaAs were also measured in a similar fashion, except the wafers were not thinned (GaAs and GaAlAs are transparent to the 1.06 µm radiation). The SiC (on axis) revealed a large stress when compared to the SiC(off axis). This difference in the amount of stress may be associated with the reduction in antiphase boundaries in SiC grown on off axis Si [5]. K. Shibahara and coworkers were able to show that the presence of anti-phase boundaries was directly related to the step height during growth. The step height associated with the (100) direction is

Table 1 Semiconductor Stress

Material	Total crystal stress
GaAs	0.30×10^5 dyne/cm^2
SiC (on axis)	1.31×10^9 dyne/cm^2
SiC (off axis)	1.69×10^8 dyne/cm^2
AlGaAs (.05)	2.72×10^6 dyne/cm^2
AlGaAs (.15)	9.28×10^6 dyne/cm^2
AlGaAs (.30)	1.23×10^6 dyne/cm^2

clearly odd; however, for off axis material, the step height is even. This even step height reduces the possibility of anti-phase boundaries, thus reducing the total amount of stress which is associated with the overall amount of birefringence. Additional work must be done in this area.

The total stress detected in the GaAs and GaAlAs/GaAs layers was much less than in the SiC, indicating that the GaAlAs/GaAs system is "almost" lattice matched, thereby reducing the total stress in the wafer. Additional research is under-way to fine tune the technique and provide a two-dimensional stress map of an entire wafer.

Conclusion

A technique for measuring interfacial and bulk stress in semiconductor materials has been developed. The technique involves the detection and quantitative measurement of birefringence in SiC, GaAs and AlGaAs systems, using a photoelastic modulator and a detector. Initial results indicate that measurements can be obtained for strains as small as those observed in a GaAs/AlGaAs interface. Although the investigators limited the sample material to semiconductors, this novel technique can be readily applied to measure stress/strain in any material system which is transparent to the selected laser radiation.

Acknowledgement

This work is supported by NASA Lewis Research Center (NAG 3-431), the National Science Foundation (RII-8413805), National Science Foundation (RII-8714767), and Pacific Telesis.

References

1. A. Kuske and G. Robertson, *Photoelastic Stress Analysis* (Wiley, New York 1974)
2. M. Born and E. Wolf, *Principles of Optics*, 6th ed. (Pergamon Press, Oxford 1980), 703.
3. J. C. Kemp, J. Opt. Soc. Am. 59, 950 (1969)
4. G.J. Felton, M.Eng. Thesis, Howard University, August 1986
5. K. Shibahara, S. Nishino, and H. Matsunami, Mat. Res. Soc. Symp. Proc. 97, 183 (1987)

Effect of Surface Modifications of Cubic SiC on Metallization Interactions

M.V. Zeller[1] *and J.J. Bellina, Jr.*[2]

[1] University of Notre Dame, Notre Dame, IN 46556, USA
[2] St. Mary's College, Notre Dame, IN 46556, USA

In order to achieve good adhesion and stable electrical properties of metal contacts on 3C-SiC, we have examined the surface chemistry of SiC single crystals after various modification treatments. All SiC samples were obtained from NASA/Lewis Research Center. The crystals were 5-10μm thick and still mounted on its growth substrate of (100)Si. The SiC surfaces as received showed several monolayers of carbonous material as well as a thin layer of oxidized Si. The surface cleaning and other preparations were carried out in situ in ultra-high vacuum. From Auger Electron Spectroscopy (AES) line shape changes of the Si and C transitions, carbon rich or silicon rich surfaces could be detected. These surface conditions were achieved by annealing or ion bombardment. (1) Ion bombardment by various rare gas ions created a disordered surface as observed by LEED (Low Energy Electron Diffraction). A combination of thermal treatment plus ion bombardment produced the (100) SiC surface. Other investigators (2) have shown annealing in a silicon flux generates an ordered surface on which the Si enrichment can be controlled.

The metal-Si interface has also been studied as a function of these surface treatments. In this presentation, we will review these results. Such metals as Ti, Ta and Cr have been found to form metal carbides on a carbon rich surface. Ni tends to form islands or diffuse into the SiC leaving excess C at the interface and yielding poor electrical properties. Detailed investigation of Ti on SiC will be discussed in this presentation. These results (3) show that initially TiC forms on SiC surface with excess C. Long term thermal treatment at 800°C causes an Si rich layer to form, with a decrease in TiC at the near surface region. Other materials to be used as high temperature contacts on SiC are also being examined in our laboratory.

1). J. J. Bellina, Jr., J. Ferrante, and M. V. Zeller, J. Vac. Sci. Technol. A4(3), 1986, p. 1692.

2). T. M. Parrill and Y. W. Chung, to be presented at the 1987 Fall AVS meeting in Anaheim, CA.

3)
J. J. Belling, Jr. and M. V. Zeller, to be published in Proceedings of the 1987 Spring MRS Meeting.

Amorphous Silicon Carbide Thin Films Produced in the Glow Discharge Deposition System

N.T. Tran

Applied Technology Laboratory, 3M Center, St. Paul, MN 55144, USA

Thin films of amorphous silicon carbide (a-SiC:H) were produced by the decomposition of methane and silane in an RF glow discharge deposition system. Optimized p^+ and n^+ a-SiC:H films with a band gap of 1.9 eV, and an activation energy of 0.4 eV and 0.23 eV, respectively, were made. The band gap of undoped a-SiC:H films could be varied from 1.9 eV to 2.7 eV by increasing the CH_4/SiH_4 gas ratio. Good quality films were obtained for a deposition rate as high as 13 Å/sec. Optical emission spectra were measured during deposition, and the emission intensities are correlated with the optical band gap of the films.

Thin film devices, including amorphous silicon solar cells, photodiodes, and light emitting diodes, that include one or more a-SiC:H layers were fabricated. The characteristics of these devices indicate that a-SiC:H films are good candidates for device applications.

INTRODUCTION

Amorphous silicon carbide films, an example of a tetrahedrally bonded amorphous alloy, are potentially useful in a large variety of device applications. The successful incorporation of a-SiC:H as a window layer in an amorphous silicon solar cell led to great improvements in conversion efficiency [1] and initiated the use of a-SiC:H in both electrical and opto-electronic devices. Hamakawa and his coworkers [2] have succeeded in obtaining emission at red, orange, yellow, and green wavelengths in a-SiC light emitting diodes (LED). These devices exhibit significant advantages over conventional crystal LED's including wide area deposition, ease of manufacturing, and low cost. Amorphous SiC:H films are easily grown by the plasma deposition of silane and hydrocarbon gas mixtures [3-4].

In this paper, we report a series of experiments on the electrical and optical properties of a-SiC:H films that were prepared by the decomposition of methane (CH_4) and silane (SiH_4) gases. Solar cells, photodiodes, and LED's were fabricated with a-SiC:H layers, and the characterization of the devices indicates that they are potentially good for applications.

EXPERIMENTAL

The deposition of a-SiC:H films from silane and methane was performed in an RF glow discharge system. Good quality films were obtained when the substrate temperature was varied between 200 - 250°C, and when RF power and pressure were varied from 5 - 15 watts and 0.5 - 1.0 Torr, respectively. For the intrinsic films, SiH_4 and CH_4 (Matheson grade) were used. In the doped films, B_2H_6 mixed in H_2 or He was used for p^+ material, and PH_3 in H_2 or He was used for n^+ material. Optical emission spectra during deposition were measured using a PAR/EG7G model 1451 optical multichannel analyzer, and the emission intensities of $SiH^*(I_{SiH})$, $H^*(I_H)$ and $CH^*(I_{CH})$ were obtained. The details of the deposition and emission spectra experiments are described in more detail in Ref. 5-6.

Springer Proceedings in Physics, Vol. 34 Amorphous and Crystalline Silicon Carbide
Editors: G.L. Harris and C.Y.-W. Yang

TABLE 1

OPTICAL AND ELECTRICAL PROPERTIES OF a-SiC:H FILMS

- Band gap from 1.9 eV to 2.7 eV for undoped a-SiC:H films.
- A deposition rate as high as 13 Å/sec for good quality films.
- A band gap of 1.9 eV, activation energy of 0.4 eV, dark conductivity of $5 \times 10^{-5} (ohm\text{-}cm)^{-1}$ for p+ a-SiC:H films.
- A band gap of 1.9 eV, activation energy of 0.23 eV, dark conductivity of 8×10^{-3} $(ohm\text{-}cm)^{-1}$ for n+ a-SiC:H films.
- At a given condition, phosphorus doping is more efficient than boron doping in a-SiC:H films.
- At a hydrogen/methane ratio of 1/20 - 1/50, and a silane/methane ratio of 20 to 30, the light/dark current ratio is 10^5 for the band gap of 1.9 eV.

RESULTS AND DISCUSSION

The electrical and optical properties of our a-SiC:H films have been published elsewhere [5-6]. A summary of these properties for our typical films is shown in Table 1.

Figure 1 shows a typical spectrum from a silane-methane plasma at 40% methane fraction in the wavelength region between 210 and 750 nm. The RF discharge power was 15 W (0.09 W/cm^2). The emission lines at 288.1, 414.2, 431.4 and 655.6 nm are assigned as those from Si^*, SiH^*, CH^*, and H^*, respectively.The optical band gap, which varies from 1.9 eV to 2.7 eV, increases as the methane fraction increases and the ratio of I_{SiH}/I_H decreases, as shown in Fig. 2. Figure 3 shows the emission intensity, I_{SiH} and I_{CH}, as a function of methane fraction. I_{SiH} decreases with increasing methane fraction as might be expected, but I_{CH} increases and then levels off in the neighborhood of 60% methane fraction. The IR vibrational spectra of a-SiC:H films with different methane fraction are shown in Fig.4.Thermodynamic calculations of the equilibrium composition of a methane-silane mixture at 3000K and 1 Torr were made. Distribution of SiC, SiC_2, CH, CH_2, SiH, and SiH_2, as a function of methane fraction, are summarized in Fig.5. Composition of the films was analyzed by x-ray Photoelectron Spectroscopy (ESCA). The results of this analysis indicate that C/Si is linearly proportional to CH_4/SiH_4 gas ratios, as shown in Fig.6.

From Figs. 2 - 6, we have observed the following:

(i) The number of CH bonds is relatively constant above 60% methane fraction, as shown in Figs. 3,4 and 5. In Fig.4, the absorption band between 2800 and 3000 cm^{-1} is the CH_n stretching mode, whereas the absorption band between 1200 and 1400 cm^{-1} is the CH_n bending mode. The band for the two CH_n modes becomes saturated at above 50% methane fraction. In Fig.5, concentration (corresponding to mole fraction) of CH and CH_2 exhibit similar behavior. If CH corresponds to the decomposition rate of CH_4 gas, then at above 60% methane fraction, decomposition of CH_4 becomes saturated.

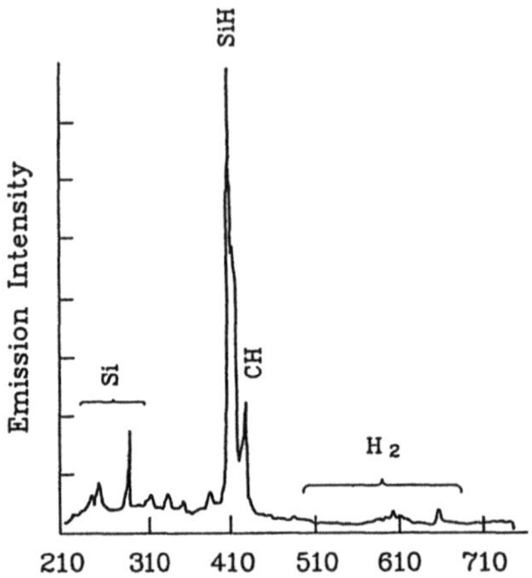

Figure 1. A typical emission spectrum from a glow discharge of silane-methane gas mixture at 1 Torr total pressure.

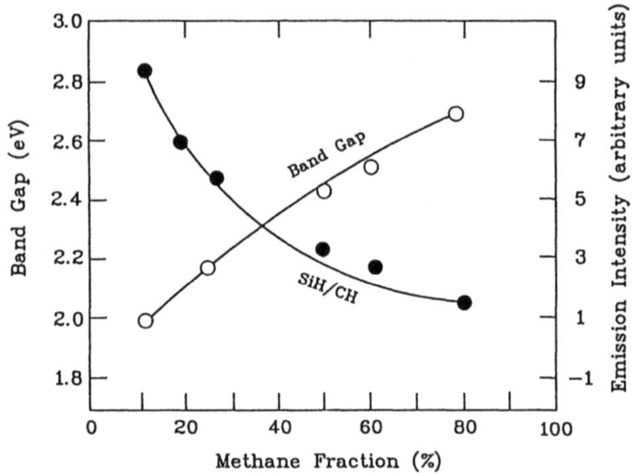

Figure 2. Variation of band gap and emission intensity ratio of I_{SiH}/I_{CH} with methane fraction.

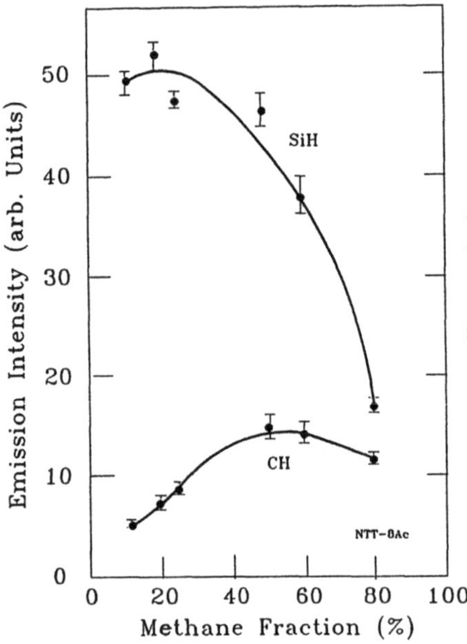

Figure 3. Emission intensity I_{SiH}, I_{CH} as a function of methane fraction.

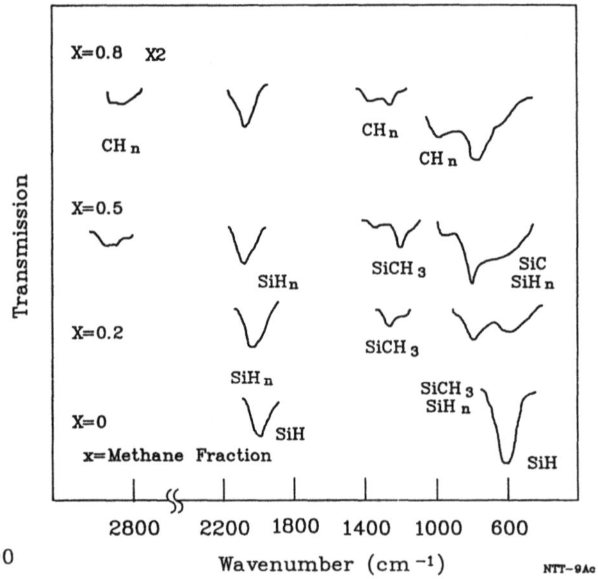

Figure 4. Infrared spectra of a-SiC:H films with different methane fraction.

This may be due to the fact that the amount of hydrogen which acts as a catalyst for CH_4 decomposition is small. The amount of hydrogen derived from the decomposition of SiH_4 gas is small at high methane fraction (Fig.3).

(ii) Carbon promotes silicon dihydride bonds in amorphous networks. In high qual amorphous silicon (a-Si:H) films, only absorption bands at 630 cm^{-1} and at 2000 cm^{-1} were observed. The two absorption bands are assigned to the rocking mode and the

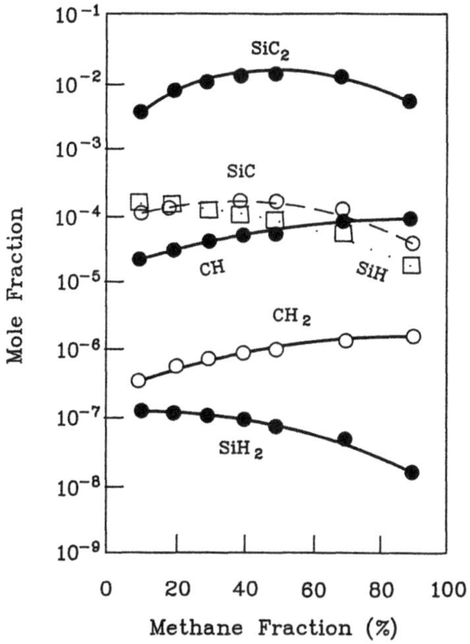

Figure 5. Calculated equilibrium concentration of SiC, SiC_2, SiH, SiH_2, CH, CH_2 as a function of methane fraction.

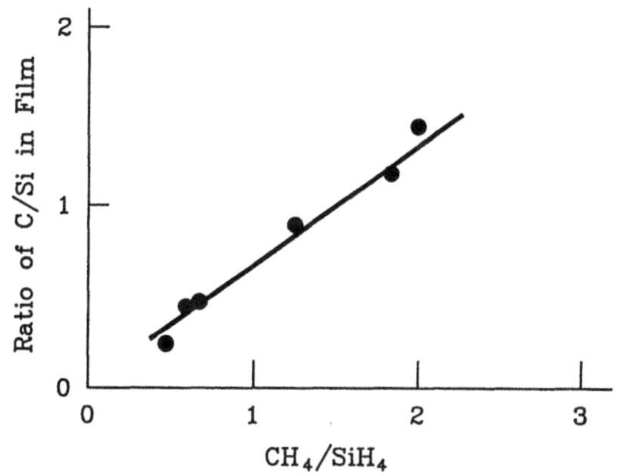

Figure 6. Relation between Si/C in the films and SiH_4/CH_4.

stretching mode of SiH bonds, respectively. With the addition of carbon, the absorption band between 845 cm^{-1} and 890 cm^{-1} was observed; in addition, the band at 2000 cm^{-1}shifted to 2090 cm^{-1}. These two absorption bands are assigned to the wagging mode and stretching mode of SiH_2 bonds.

(iii) As shown in Fig.4, in a-SiC:H films obtained from the RF decomposition of methane and silane, carbons are mostly incorporated as methyl (CH_n) groups.

(iv) SiH/CH ratio is a good indicator (or reference) for controlling the band gap in a-SiC:H films, as shown in Fig.2.

(v) Dissociation rate of CH_4 might be smaller than that of SiH_4, as shown in Figs. 3 and 6.

APPLICATIONS

Amorphous silicon carbide films of intrinsic, p^+, and n^+ material were used in the fabrication of a variety of devices. The materials had properties comparable to the properties of the films discussed in the previous sections. Solar cells, photodiodes and light emitting diodes were all fabricated and the characteristics of each type of device are discussed below.

- SOLAR CELLS

p^+ a-SiC:H film was used as a window layer in amorphous silicon solar cells and the effects of carbon grading at the interface between p^+ a-SiC:H and intrinsic a-SiC:H were studied. Fig. 7 shows the dependence of open circuit voltage V_{oc} on the carbon profile at the buffer layer of p-i-n and n-i-p amorphous silicon solar cells which has $\lambda/4$ Indium Tin Oxide (ITO) electrode. For the p-i-n solar cells, V_{oc} increased from 0.73 eV to 0.86 eV as the profile depth varied from 0 to 400 Å. For the n-i-p solar cells, a similar trend was observed; however, the increase of V_{oc} with carbon profile

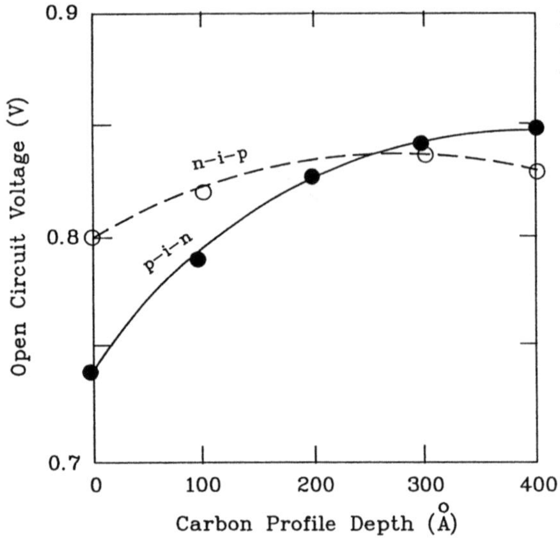

Figure 7. Dependence of V_{oc} on the carbon profile at the buffer layer of p-i-n and n-i-p amorphous silicon solar cells.

depth was slow. In the p-i-n amorphous silicon solar cells on polyimide substrate, because the p^+ a-SiC:H layer was deposited last, an abrupt change of composition between intrinsic a-Si:H and p^+ a-SiC:H layers could induce a large number of defects at the p/i interface. As a result, the interface recombination current will increase. Since, by definition, open circuit voltage is the point at which the photogenerated current equals the sum of the forward bias diffusion and the recombination current, an increase in recombination current will therefore reduce V_{OC} by using such a p-i-n solar cells on polyimide substrate, in order to increase V_{OC} by using such a window layer with larger optical band gap such as p^+ a-SiC:H, a buffer layer which has a carbon compositional grading between p^+ a-SiC:H and a-Si:H is required. Further details were discussed elsewhere [7-8].

- PHOTODIODES

A photodiode was formed from a heterojunction of p^+ a-SiC:H and undoped a-Si:H. In this heterojunction diode, 200 - 300 Å thick p^+ a-SiC:H deposition followed 1 μm thick a-Si:H deposition. The I-V characteristics of a typical device in the dark and under illumination with 260 μW/cm^2 of 540 nm light are shown in Fig.8. It should be noted that these characteristics are for photodiodes without any antireflection coating; the photocurrent is therefore expected to improve more with antireflection coating such as ITO. The diodes exhibit very small dark reverse current (10^{-13} A/cm^2

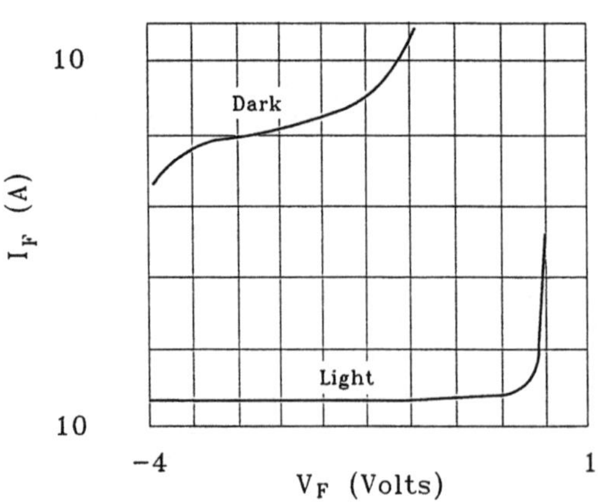

Figure 8. I-V characteristics of a-SiC:H/a-Si:H heterojunction photodiode.

at -0.5 V) , and photo to dark current ratio as high as 10^5 has been obtained. The photocurrent saturates without any applied voltage. The high built-in potential, formed between p^+ a-SiC:H and undoped a-Si:H, fully depletes undoped a-Si:H layer. Photogenerated carriers drift toward each electrode due to the high internal field in the undoped a-Si:H layer, which then makes the photocurrent saturate without applied voltage. The low dark current and high photosensitivity of the heterojunction diode make it a good candidate for application as an image sensor.

-LIGHT EMITTING DIODES (LED's)

Amorphous SiC thin film light emitting diodes of the Hamakawa-type were studied. LED's have a structure of p^+ a-SiC:H/ undoped a-SiC:H/ n^+ a-SiC:H. The thickness of p^+ and n^+ a-SiC:H layers was in the range of 150 Å to 400 Å . The cell area was 0.01 cm^2. The band gap of the undoped (intrinsic) a-SiC:H layer varied from 2.3 eV to 2.7 eV. An undoped layer with a band gap of 2.7 eV exhibits a photoluminescence peak at 481 nm in the blue region of the visible spectrum, as shown in Fig.9. This peak corresponds with the photoluminescence peak of single crystal SiC for blue LED's [9]. The results support the fact that amorphous SiC blue LED's are possible. The I-V characteristic of one a-SiC:H p-i-n diode is shown in Fig.10. The I-V characteristic of a crystalline SiC blue LED is shown in the inset of Fig. 10 for comparison. The amorphous SiC:H p-i-n diode exhibits a threshold voltage of 22 V, whereas the crystalline SiC diode has a threshold voltage of 2 V. This threshold voltage (22V) is still far lower than that of other types of a-SiC LED's and ZnS LED's which are more than 100 V [10]. The difference in threshold voltage between amorphous and crystalline SiC devices can be narrowed by controlling the band gap and the thickness of the undoped a-SiC:H layer, as can be seen in Fig. 11. The threshold voltage, the voltage at which the current sharply increases, depends partly on the series resistance in the undoped a-SiC:H layer and partly on the barrier height at both p/i and i/n interface. In this type of device, p^+ a-SiC:H and n^+ a-SiC:H play the role of the injector of holes and electrons into the luminescence-active undoped layer. Although the luminescence peak of the undoped a-SiC:H is in the blue region, the LED's produced from a-SiC:H (electroluminescence mode) emit intermittently light that is either orange or white green. The lower energy of electroluminescent emission presumably results from the fact that electroluminescence takes place at the p/i interface, where the band gap is graded from 2.7 eV (undoped layer) to 1.9 eV (p^+ layer). Therefore, in order to obtain a-SiC:H blue LED's, improvement in the large band gap p^+ a-SiC:H is required.

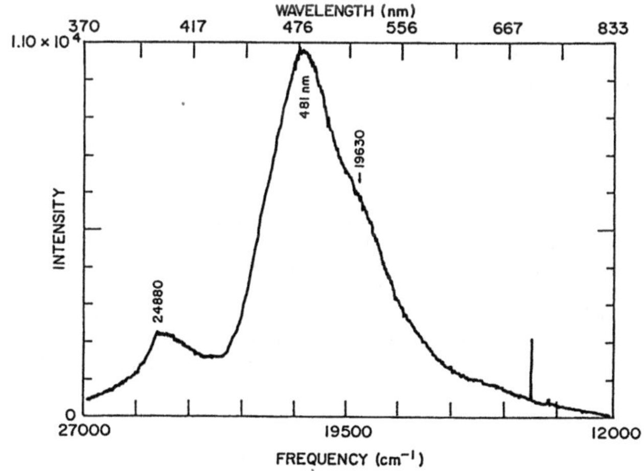

Figure 9. Photoluminescence spectrum of our a-SiC:H films (room temperature).

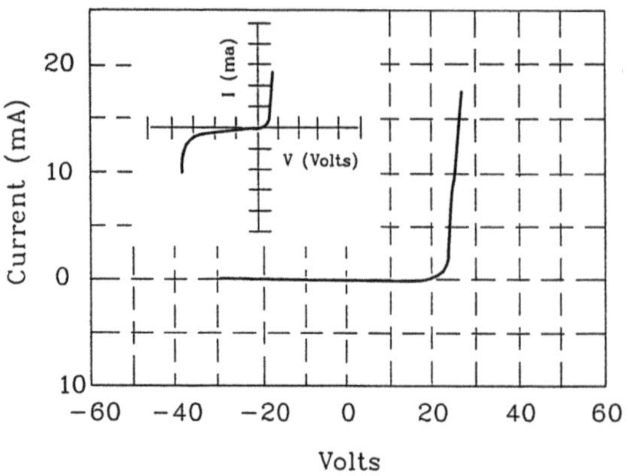

Figure 10. I-V characteristics of a-SiC:H p-i-n thin film diodes.

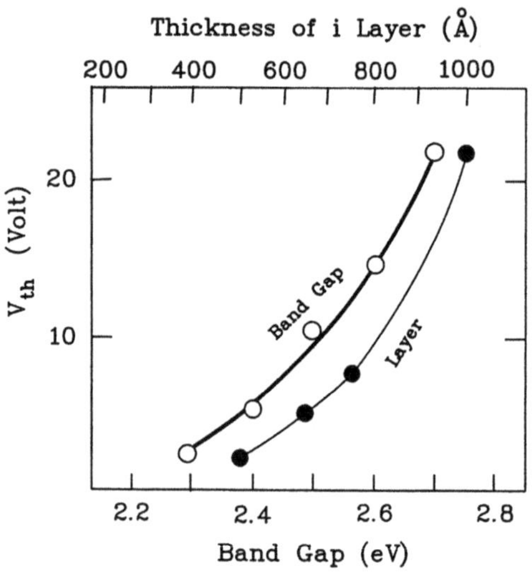

Figure 11. Variation of the threshold voltage voltage (V_{th}) as a function of band gap and thickness of the undoped a-SiC:H layer.

CONCLUSIONS

Good quality amorphous silicon carbide thin films were produced by the decomposition of methane and silane in an RF glow discharge deposition system.

High quality solar cells and photodiodes, with the incorporation of one or more a-SiC:H layers, were successfully demonstrated. However, for stable, blue LED's, a better quality for high band gap p^+ a-SiC:H is required. The material can be prepared by improving the current RF diode glow discharge system or by utilizing other deposition techniques, such as triode or electron cyclotron resonance chemical vapor deposition.

ACKNOWLEDGMENTS

The author would like to thank D. Olsen, R. Patel, F. Jeffrey, F. Mori, K. Epstein, and D. Misemer for helpful discussions, suggestions and technical assistance during the course of this work.

REFERENCES

1. Y. Hamakawa and Y. Tawada, Int. J. Solar Energy 1, 125 (1982).
2. D. Kruangam, T. Endo, M. Deguchi, W. Guang-Pu, H. Okamoto, and Y. Hamakawa, Optoelectronics-Devices and Technologies, (Mita Press), vol. 1, no. 1, 67 (1986).
3. D.A. Anderson, and W. E. Spear, Phil. Mag. 35, 1 (1977).
4. Y. Catherine, G. Turban, and B. Grolleau, Thin Solid Films 23, 765 (1981), also, Y. Catherine, Ph.D. Thesis, Université de Nantes, 1981.
5. N. T. Tran, D. J. OLsen, R. I. Patel, and K. A. Epstein, Materials Research Society Symp. Proceedings, vol. 70, 325 (1986).
6. R. I. Patel, J. Schirk, D. J. Olsen, and N. T. Tran, ibid., page 55.
7. N. T. Tran, and F. R. Jeffrey, Phys. Stat., Sol. (a) 99, K143 (1987).
8. N. T. Tran, F. R. Jeffrey, and D. J. Olsen, Materials Research Society Symp. Proceedings, Anaheim, California, April 1987 (to be published).
9. H. Matsunami, M. Ikeda, A. Suzuki, and T. Tanaka, IEEE, ED-24, no.7, July, 958 (1977).
10. H. Matsunami, M. Yoshimoto, Y. Fujii, and J. Saraie, J.Non-Cryst. Solids 59&60, 569 (1983).

Preparation and Properties of Polycrystalline Silicon Carbide Films Produced by Plasma Enhanced Chemical Vapor Deposition, and Their Applications

Y. Onuma, F. Nagaune, and K. Kamimura

Department of Electrical Engineering, Shinshu University,
500 Wakasato, Nagano-shi 380, Japan

Polycrystalline SiC films were prepared by plasma enhanced chemical vapor deposition (plasma-CVD), and characterized by X-ray diffraction, photo-absorption and Hall measurement. The piezoresistive gauge factor (sensitivity of piezoresistive sensor) of the film was between 5 and 10. A polycrystalline SiC/single crystal Si contact had sensitivity to the light of wavelength less than 450 nm.

Introduction

Silicon carbide has attracted increased attention because of its wide energy gap and mechanical hardness. Much work[1-6] has been reported about epitaxial films or amorphous films and their applications to electronic devices such as heterojunction bipolar transistors, Schottky gate field-effect transistors and solar cells. In these studies, well-established methods have been used for the preparation of the SiC films, but a rather simple method is useful for the preparation of polycrystalline films which are used in some devices such as piezoresistive sensors.

One of the limiting factors for the application of crystalline SiC films is the high substrate temperature during deposition. The substrate temperature must be higher than 1000 °C to obtain an epitaxial SiC film. The plasma-assisted process is effective to reduce the process temperature.

This paper dealt with the preparation, characterization and application of polycrystalline SiC films by plasma enhanced chemical vapor deposition.

Experiment

Polycrystalline SiC films were prepared by plasma enhanced chemical vapor deposition (plasma-CVD). Figure 1 shows the schematic representation of the reaction chamber. The chamber was made from stainless steel. The source gases were introduced to the chamber through the multi-pore gas entrance, which acted as an electrode for the glow discharge. This structure of the electrode made it possible for the source gases to be excited effectively by the radio frequency (RF) electric power. A mixture of 10 % SiH_4 in H_2 and a mixture of 10 % CH_4 in H_2 were used as the source gases. The flow rates of these gases were controlled by float meters and needle valves. The ratio of the flow rate of SiH_4 to that of CH_4 was fixed to 0.1 throughout this work. The substrate holder was made from a graphite block. This holder acted also as a substrate heater. The substrate temperature was set from 600 to 900 °C by controlling the current in this block. The substrate

Springer Proceedings in Physics, Vol. 34 Amorphous and Crystalline Silicon Carbide
Editors: G.L. Harris and C.Y.-W. Yang

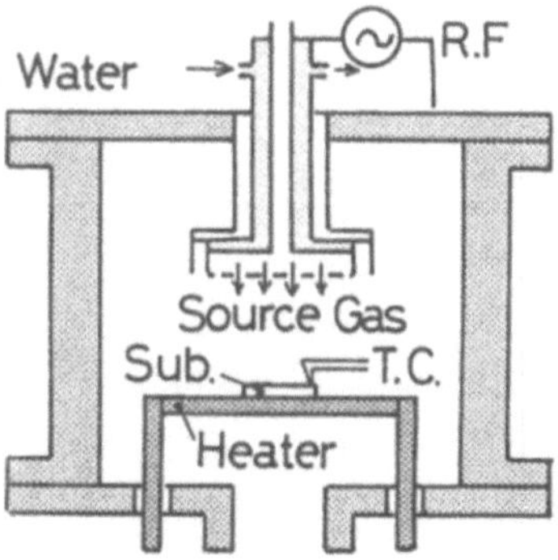

Fig. 1 Schematic diagram of the reaction chamber

temperature was measured by an alumel-chromel thermocouple attached to the substrate surface. The system was evacuated by a conventional oil rotary pump and the pressure in the chamber was measured by a Pirani gauge. The pressure was about 1 Torr during the film deposition. Thermal oxdized SiO_2 films and fused quartz glasses were used as substrates.

The film thickness was determined from the observation of the cross section of films by a scanning electron microscope. X-ray diffraction analysis was carried out to characterize the structure of the films. The electrical properties were determined from Hall measurements. The piezoresistive gauge factor was estimated to show the promising characteristics for the application to a strain gauge and pressure sensor. The measured values were plotted against the substrate temperature. The other process parameters were fixed to constant values throughout this work. The SiC/Si contacts were prepared and photovoltaic characteristics were measured.

Results and Discussions

The deposition rate was determined from the film thickness. Figure 2 shows the dependence of the deposition rate on the substrate temperature together with the data reported in preceding works[1,2]. The slope of the linear lines for this work in the low temperature range in Fig. 2 seems almost the same as that of the previous works. This means that the deposition process is limited by a thermal reaction on the substrate surface even in plasma-CVD at a low temperature range. The thermal reaction at the surface of the substrate may give rise to the temperature dependence of the deposition rate. It is apparent that the plasma process is effective for increasing the deposition rate in the low temperature range. When the substrate temperature was higher than 700 °C, the deposition rate was almost constant. This may be because the deposition rate is limited by the transport of plasma-induced radicals to the substrate surface.

Figure 3 shows the dependence of the X-ray diffraction pattern on the substrate temperature. The diffraction peaks became larger and sharper as the substrate temperature increased, except for the data of the sample prepared at 900°C. The grain size increases as the substrate temperature increases. The film thickness decreases at the substrate temperature of 900°C, and this may be the reason for the decrease in the diffraction peaks for the sample prepared at 900°C. The grain size was estimated from the full width at half maximum of the highest peaks in each pattern. The grain size increased from 15 to 30nm as the substrate temperature increased from 650 to 900°C.

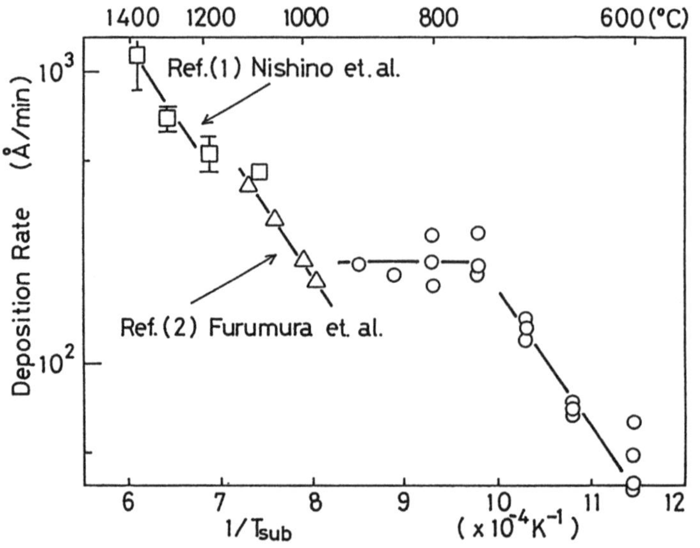

Fig. 2 Temperature dependence of deposition rate

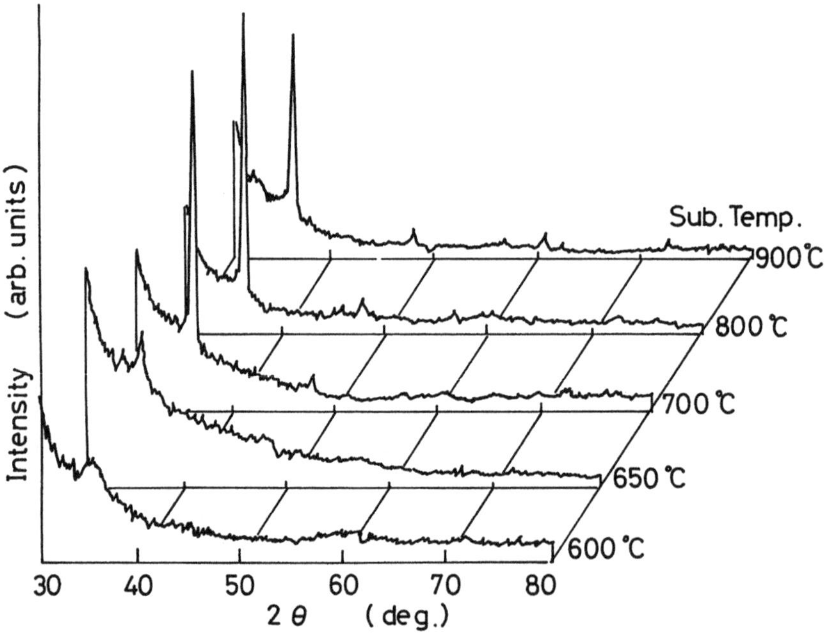

Fig. 3 Dependence of X-ray diffraction pattterns on substrate temperature

The crystal structure of the film cannot be identified from the results in Fig.3 because almost all phases of SiC contain the atomic spacing corresponding to the diffraction near 35 degrees. The band gap energy depends on the crystal structure. The band gap energy was estimated from the photo-absorption measurement to identify the crystal structure of the films. Figure 4 shows the spectral response of the absorption coefficient of the polycrystalline SiC films. The absorption coefficents were estimated

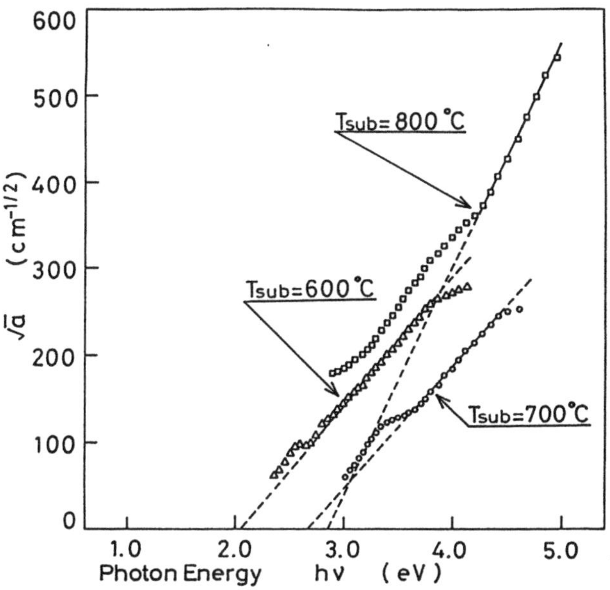

Fig. 4 Spectral response of absorption coefficient

from the decay of the light intensity through the films and the thickness of the films by neglecting the reflection at the surface. The square root of the absorption coefficients were plotted against the energy of the incident light in Fig.4, and the linear portions were extrapolated to zero absorption. The silicon carbide has indirect band structure with an energy gap of 2.20 eV for 3C structure and of 2.86 eV for 6H structure. The film prepared at 800 °C seems to consist mainly of hexagonal SiC, because the absorption edge is about 2.8 eV.

The lower energy values of the absorption edge were obtained for the samples prepared at the lower substrate temperature. The excess Si was observed in the films prepared at low substrate temperature (less than 700°C) or with low incident RF power (less than 50 W) by the electron probe micro analysis (EPMA) measurement. For some of these samples, diffraction peaks corresponding to crystalline Si were observed in X-ray diffraction patterns. This may be one of the causes of the low absorption edge of the sample prepared at 600 °C. The position of the diffraction peaks near 35 degrees were not affected by the change in substrate temperature or RF power, even if the peaks corresponding to Si were observed. The film consists of a mixture of Si and SiC when it is prepared at low temperature or with low RF power.

The electrical characteristics were evaluated by Hall measurement. Figure 5 shows the dependence of the carrier concentration and Hall mobility on the substrate temperature. All films showed n-type conduction. No impurity was doped intentionally and the origin of the donor was not identified. The Hall mobility was about 10 cm^2/Vs and was almost independent of the substrate temperature. The carrier concentration ranged from the order of 10^{16} to 10^{18} cm^{-3}. A definite relation was not observed between the carrier concentration and the substrate temperature. The resistivity ranged from the order of 10^0 to 10^1 ohmcm.

The polycrystalline SiC film should be one of the suitable materials for pressure sensors used at high temperature. The piezoresistive effect of the polycrystalline SiC films was measured to show the promising properties for the application as a pressure sensor. Figure 6 shows the relations between the change in

Fig. 5

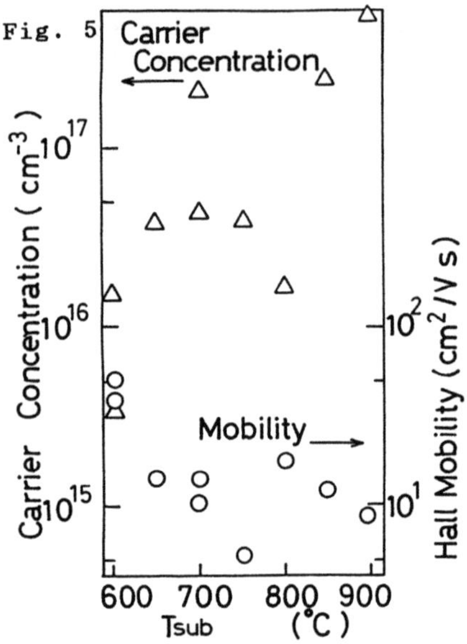

Fig. 6

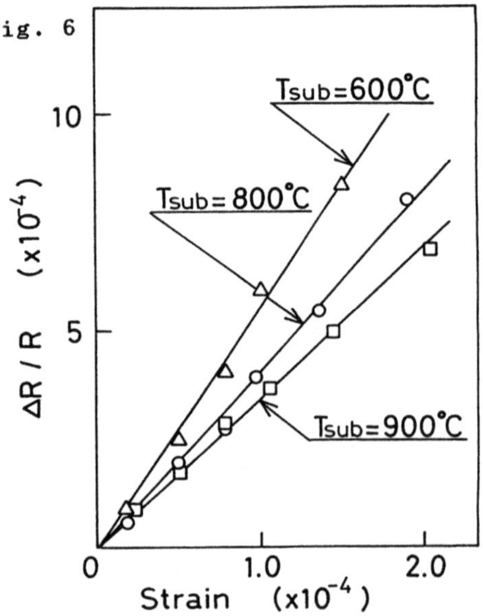

Fig. 5 Dependence of carrier concentration and Hall mobility on substrate temperature

Fig. 6 Relation between change in resistance and strain

resistance and the strain in the film. The measurements were carried out by a conventional cantilever system. The films were attached to the steel cantilever with epoxy adhesives for a strain gauge. The strain in the film was estimated from the bend of the cantilever. The resistance of the film was measured by Wheatstone bridge.

Figure 6 shows the relations between the change in resistance and the strain in the film. In all cases in this work, resistance increased when the tensile strain was applied to the film. The sensitivity of piezoresistive elements (gauge factor) ranged between 5 and 10. These values were not so large as those of other semiconducting polycrystalline films such as Ge or Si. But it is fairly large as compared to metal films.

Figure 7 shows the spectral response of short circuit current in a polycrystalline n-SiC/single crystal p-Si structure. The light was introduced to the SiC/Si interface through the SiC layer. The polycrystalline SiC layer acted as a window and photocurrent was observed even in the short wavelength (blue) region.

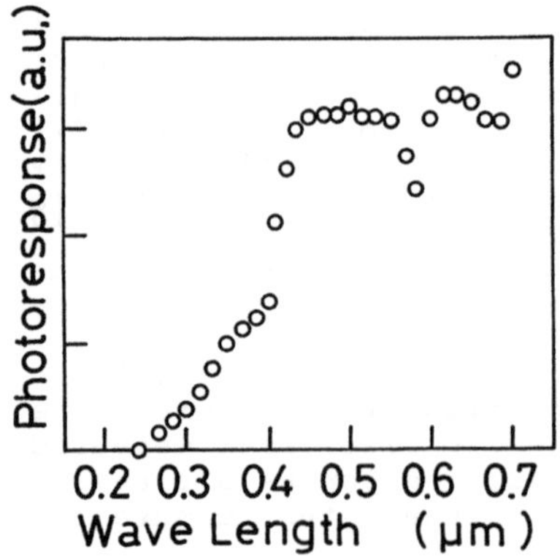

Fig. 7 Spectral response of normalized photo-current

Conclusion

Polycrystalline SiC films were prepared on insulating substrates by plasma-CVD. Properties of films depend largely on the substrate temperature during deposition. A substrate temperature higher than 700 °C was needed to obtain a high quality film. The piezoresistive gauge factor ranged between 5 and 10 and was not so large as the values for other semiconducting films. The polycrystalline SiC films may be useful for pressure sensors used at high temperature.

The polycrystalline SiC layer acts as a window in SiC/Si photodiodes.

References

1) S.Nishino,Y.Hazaki, H.Matsunami and T.Tanaka; J.Electrochem.Soc., Vol.127,(1980),pp.2674-2680.
2) Y.Furumura, M.Doki, F.Mieno and M.Maeda ;Trans. IEICE. Jpn (in Japanese), Vol.J69-C,(198),pp.705-714.
3) J.A.Freitas,Jr,S.G.Bishop,J.A.Edmond,J.Ryu and R.F.Davis;J.Appl. Phys.Vol.61,(1987),pp.2011-2016.
4) B.Segall,S.A.Alterovitz,E.J.Haugland and L.G.Matus;Appl.Phys. Letters,Vol.49,No.10,(1987),pp.584-586.
5) M.Fatemi and P.E.R.Nordquist;J. Appl. Phys.,Vol.61,(1987), pp.1883-1890.
6) S.Yoshida,H.Daimon,M.Yamanaka,E.Sakuma and K.Endo; J. Appl. Phys. Vol.60,(1986),pp.2989-2991.

Part IV

Silicon Carbide Processing and Device Applications

Study on Current Gain Degradation in Amorphous SiC Emitter HBT

K. Sasaki, T. Fukazawa, and S. Furukawa

Department of Applied Electronic, Graduate School of Science & Engineering, Tokyo Institute of Technology, 4259 Nagatsuda, Midori-ku, Yokohama 227, Japan

1. INTRODUCTION

The heterojunction emitter configuration is strongly required to overcome material restrictions and difficulties in the fabrication process of present homojunction transistors. In this situation, amorphous [1] and crystalline SiC [2] are being investigated as hetero-emitter materials for Si heterojunction bipolar transistors (HBT's). We have mainly employed and developed amorphous SiC films for Si-HBT for the following reasons:

1)The amorphous SiC hetero-emitter has complete compatibility with the poly-Si emitter which is used for high speed self-aligned bipolar transistors.
2)Not only can high injection efficiency be obtained, but also the low temperature fabrication process of amorphous SiC films makes it possible to fabricate a heavily doped and very thin base layer.
3)High temperature processes after the deposition of amorphous SiC are not restricted, since amorphous SiC has good thermo-stability.
4)It is useful for band profile design that amorphous SiC has band gap changeability.

In spite of the above promising potentialities of amorphous SiC, we have not always been able to obtain satisfactory results. In this paper, we will discuss distinctive problems in pnp and npn type amorphous SiC emitter HBT's, considering the current transport mechanisms based on the band structure model and leakage current flowing via localized states in the amorphous SiC emitter. In order to verify the current gain degradation mechanisms, an example using micro-crystalline Si and realizing high current gain will be given.

2. EXPERIMENTAL PROCEDURE

2.1 Device Fabrication

For the fabrication of the transistors, a conventional Si IC process was used. Fig.1 illustrates the cross-sectional view of the devices.

A field oxide film was grown thermally on a chemically precleaned Si substrate, and a window was cut into the oxide film. Ion implantation was carried out for fabricating the base region. A second oxidation and activation anneal for implanted impurities were carried out. After an emitter window was opened, the

Springer Proceedings in Physics, Vol. 34 **Amorphous and Crystalline Silicon Carbide**
Editors: G.L. Harris and C.Y.-W. Yang

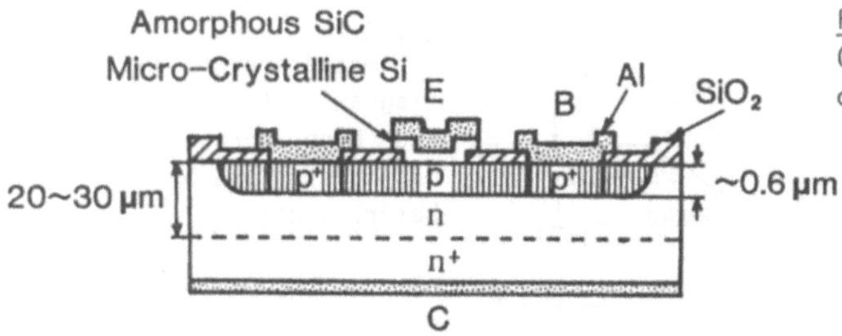

Fig.1.
Cross-sectional view of an npn type Si HBT.

base surface, at which the hetero-interface would be located, was cleaned again by RCA boiling. Immediately after dipping in HF solution to remove the chemical oxide, the samples were loaded into the discharge chamber of a plasma CVD system wherein p and n type amorphous SiC or n type micro-crystalline Si film were deposited. Emitters were patterned by CF_4 RIE method, and Al electrodes were fabricated.

2.2 Film Preparation

Amorphous SiC and micro-crystalline Si films are deposited by using an inductive coupled plasma CVD system. The deposition conditions and fundamental properties are shown in Table 1.

3. RESULTS AND DISCUSSION

3.1 Band discontinuity factor

At an early stage of this investigation, we attempted to determine the energy band structure of the amorphous SiC/crystalline Si heterojunction. The possible band structures for pn and np heterojunctions are illustrated in Fig.2. The theory and procedure used in determining the band structures are described in the literature [3].

Table 1. Deposition conditions and fundamental properties of films.

	Micro-crystalline Si	Amorphous SiC n-type	Amorphous SiC p-type
Sub. Temp.	350 °C	350 °C	450 °C
Depo. Method R.F. Power	L-coupled PCVD 60 W		
Gas Sources	SiH_4 PH_3, 1% Ar, 20%	$SiH_4 : CH_4$= 0.8 : 0.2	
		PH_3, 3%	B_2H_6, 15%
Optical Band Gap(eV)	1.75	1.9	1.8
Conductivity (S/cm)	7.1	10^{-6}	10^{-3}

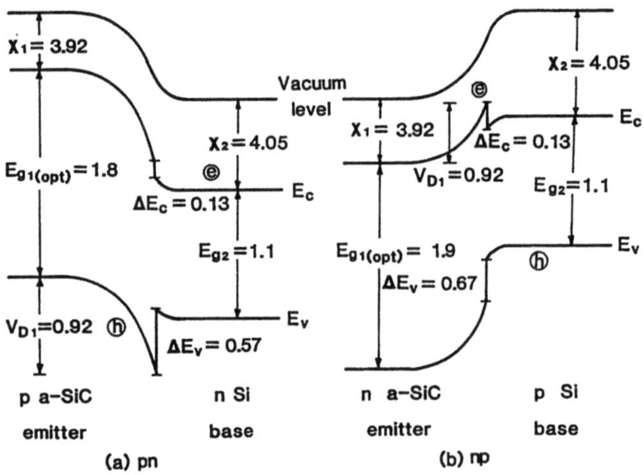

Fig.2.
Estimated energy band diagram of the (a)pn and (b)np amorphous SiC/crystalline Si heterojunction.

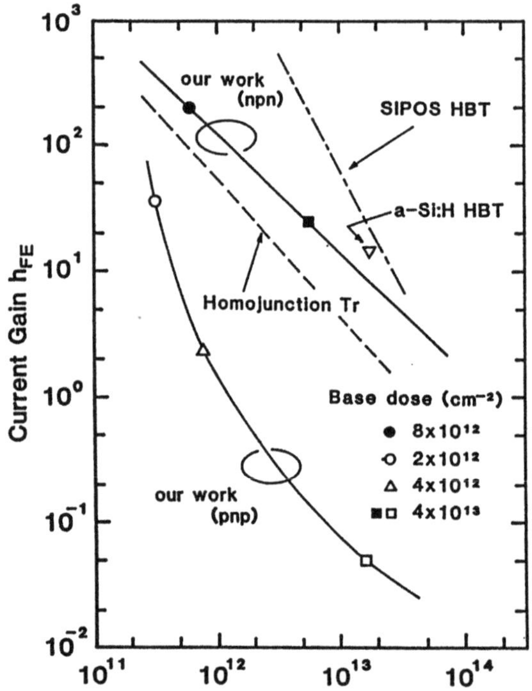

Fig.3.
Experimental results of current gain for pnp and npn HBT's. Other researchers' results are also listed as a function of base Gummel number.

According to the heterojunction theory [4], the reverse injection current from base to emitter is reduced by means of the extra barrier of ΔE_C for electrons, thus the current gain increases by the factor of $\exp(\Delta E_C/kT)$ for the pnp type HBT comparing with the homojunction transistor. However, as can be seen in the figure (a) the barrier ΔE_C is not so large. On the other hand, in npn type HBT shown in figure (b), a large barrier ΔE_V for holes exists. Therefore, higher current gain would be expected in npn type HBT's. In order to confirm this consideration, we actually fabricated pnp and npn type HBT's and measured their current gains. The results are shown in Fig.3, where current gains are plotted as a function of base

Gummel number. Obviously we can see that higher current gain is obtained in npn type HBT's.

3.2 Carrier transport by forward tunneling

In an ideal HBT, the current gain would be almost independent of base impurity concentration, because reverse injection could be neglected, therefore current gain should be determined by only base transport efficiency. Actually, however, it is observed that current gain is decreased due to various leakage currents.

As can be seen in Fig.3, the current gains of both samples of pnp and npn type HBT decrease with increasing base impurity concentration. Especially, for pnp type HBT the current gain rapidly decreases. This behavior could be explained by a current transport model, the so-called capture tunneling model [5], of an amorphous SiC/crystalline Si heterojunction [6] as follows:

A current flowing through the emitter tends to be drift-current-limited because of high resistivity of the amorphous SiC films, therefore, the current will tend to increase linearly. On the other hand, a current flowing via localized states existing in the forbidden gap of amorphous SiC, will increase exponentially because in the emitter region the electric field caused by the emitter series resistance rises in such a direction as to enhance the current. We have named this current the 'defect-assisted tunneling current'. Current gain is suppressed by such mechanisms.

We were able to find the current flow in a p-type amorphous SiC/n-type crystalline Si heterojunction. The diode n-values derived from the I-V characteristics of the heterojunction are plotted for various substrate impurity concentrations in Fig.4. From this figure, it could be explained that the degradation of n-value is due to the increase of electrons with increase of substrate impurity concentration which enhances the tunneling electron current from the conduction band of the base to localized states in the amorphous SiC emitter.

A similar phenomenon can be observed also in npn type HBT. Forward current of an emitter-base hetero-diode and current gain are plotted against the applied voltage in Fig.5. As can be seen in this figure, the current gain degradation starts at the same point at which the diode current starts saturating. At this current region the diode current is dominated by the drift current, where the electric field rises. Therefore, it could be considered that this current gain degradation is caused by the defect-assisted tunneling current mentioned above.

3.3 Control of the defect-assisted tunneling current

It can be concluded from the above investigation that the current gain degradation is due to high resistance of the emitter, in other words, low current supply capability of the emitter. If so, we will be able to obtain a high current gain by employing a low resistance emitter material. In order to confirm the validity of this interpretation, a micro-crystalline Si emitter providing wide band gap and low resistance was employed as an example [7]. The results are shown in Fig.6. A

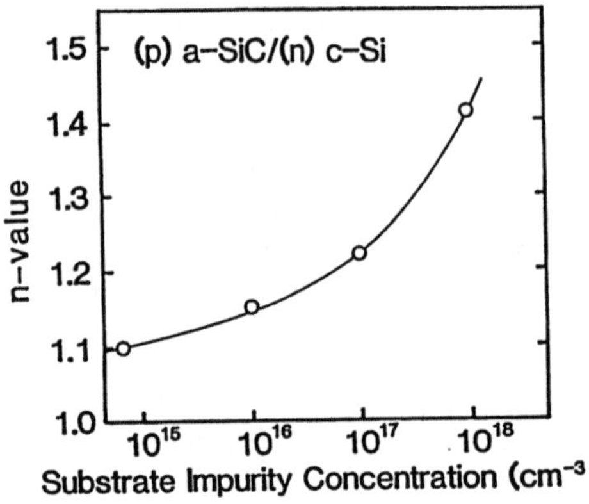

Fig.4.
Substrate impurity concentration dependences of the diode n-value.

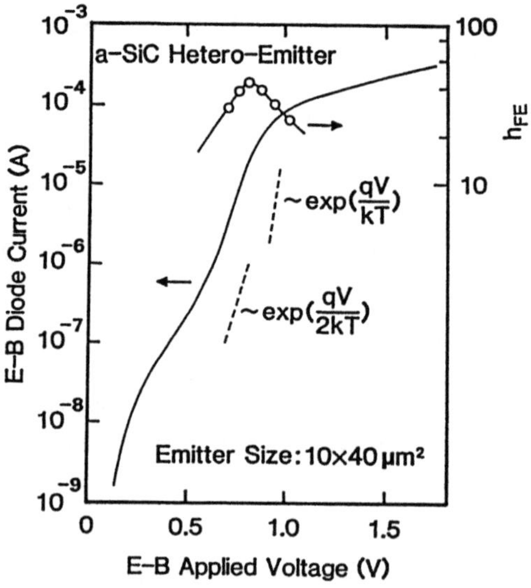

Fig.5.
Emitter-base hetero-diode forward current and current gain dependences on V_{BE}.

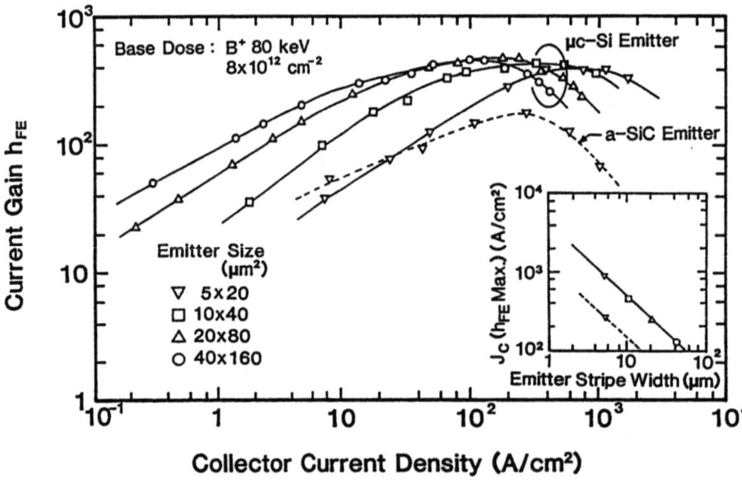

Fig.6. Current gain - collector current density characteristics and collector current density dependences on emitter stripe width of hetero-emitter transistors.

current gain as high as 480 is achieved, and this value is 2 or 3 times larger than that of an amorphous SiC emitter HBT. Furthermore, the collector current density is also improved.

4. CONCLUSION

We have discussed several factors determining the current gain of the HBT's from the viewpoint of band structure and defect-assisted tunneling current. It can be concluded that the current gain and collector current density are expected to be further improved by increasing the current supply capability. In order to achieve this condition, it is considered that the micro-crystalline phase is very useful for SiC materials.

REFERENCES

[1]K. Sasaki, S. Furukawa and M. M. Rahman : 1985 Int. Electron Devices Meeting, Washington D.C., Tech. Dig., pp.294-297.

[2]T. Sugii, T. Ito, Y. Furumura, M. Doki, F. Mieno and M. Maeda : 1986 Symp. VLSI Tech., Tech. Dig., pp.45-46.

[3]K. Sasaki, S. Furukawa and M. M. Rahman : Ext. Abs. 17th Conf. Solid State Devices and Materials, Tokyo, pp.385-388 (1985).

[4]H. Kroemer : Proc. IEEE, 70 [1], pp.13-25 (1982).

[5]H. J. Hovel and A. G. Milnes : Int. J. Electronics, 25[3], pp.201-218 (1968).

[6]K. Sasaki and S. Furukawa : Ext. Abs. 19th Conf. Solid State Devices and Materials, Tokyo, pp.335-338 (1987).

[7]K. Sasaki, T. Fukazawa and S. Furukawa : 1987 Int. Electron Devices Meeting, Washington D.C., Tech. Dig., pp.186-189.

Investigations of Stability of a-SiC Under Heat Treatment During Device Fabrication

T. Fukazawa, K. Sasaki, and S. Furukawa*

Department of Applied Electronics, Graduate School of Science & Engineering, Tokyo Institute of Technology, 4259 Nagatsuda, Midori-ku, Yokohama 227, Japan

1. INTRODUCTION

In recent years, with a view to increase the speed of Si bipolar transistors, it has become important to utilize a heterojunction bipolar transistor (HBT). We have studied the materials characteristics of amorphous silicon carbide (a-SiC) as a hetero-emitter of an HBT [1]. a-SiC has been reported to have promising features such as

1) relatively good interface characteristic with Si [1]
2) wide band gap
3) easy to form at low temperatures.

It has been shown that the a-SiC emitter HBT has higher current gain than a conventional Si homojunction transistor [2],[3].

In order to apply a-SiC to Si IC's containing HBT's, however, it is necessary to have information about the heat treatment characteristics, i.e. the stability of this film during thermal processes. This paper describes the thermal stability of a-SiC.

2. EXPERIMENTAL PROCEDURES

Phosphorus doped n-type a-SiC films were deposited by an induction coupled plasma CVD system [1]. Deposition condition of a-SiC films is shown in Table 1. Deposited a-SiC films were then annealed at 400 to 700°C for 30 min in vacuum or nitrogen/hydrogen mixed gas ambient. After annealing, measurements of dark conductivities, optical band gap and IR absorption were carried out.

Added to the above measurements, we fabricated a-SiC/c-Si heterojunction diodes and evaluated their I-V characteristics after annealing. Figure 1 shows a cross-sectional view of a-SiC/c-Si heterojunction diode. First, 80keV B^+ ions were implanted in a p/p^+ Si substrate to a dose of $8\times10^{12}cm^{-2}$. Then the sample was oxidized in wet oxygen ambient at 1025°C for 30min in order to form an SiO_2 layer and to activate implanted species. A window was cut into the oxide and the Si surface was cleaned by RCA boiling. Immediately after dipping in the HF solution, the sample was loaded into the plasma CVD system and an a-SiC film was deposited. Then the sample was annealed for 30min in nitrogen/hydrogen mixed gas ambient. Finally Al electrode was formed.

*On leave of absence from Nippondenso Co., Ltd.

Springer Proceedings in Physics, Vol. 34 Amorphous and Crystalline Silicon Carbide
Editors: G.L. Harris and C.Y.-W. Yang

Table 1 Deposition condition of a-SiC

Substrate Temp.	350 °C
Deposition Method	L-coupled PCVD
RF Power	60 W
Gas Sources	$SiH_4 : CH_4$ = 0.8 : 0.2 PH_3 3 % (He diluted)
Film Thickness	150 nm

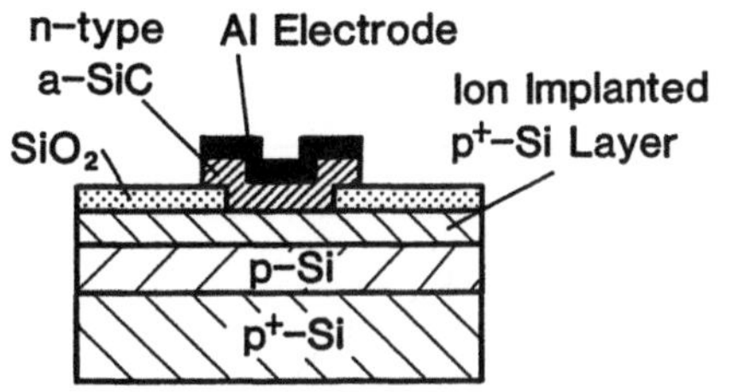

Fig.1 Cross-sectional view of a-SiC/c-Si heterojunction diode

3. RESULTS AND DISCUSSION

Figure 2 shows a relation between dark conductivities of a-SiC and annealing temperatures. Up to 600°C, the conductivity increases with annealing temperature. This is considered due to the densification of the films. However, the conductivity decreases drastically after annealing at 700°C.

The band gap which is one of the important features as a hetero-emitter of HBT, were measured next. Figure 3 shows photo-absorption spectra of a-SiC films annealed at various temperatures. Optical band gaps are indicated by the intercept of the straight line with photon energy axis in Fig.3. The dependence of optical band gaps, obtained from Fig.3, on annealing temperature is shown in Fig.4. It is observed that the decrease of optical band gaps with increasing annealing temperature is relatively small up to 600°C. After annealing at 700°C, however, optical band gap decreases drastically. These results are considered due to the dissociation of bonded hydrogen atoms.

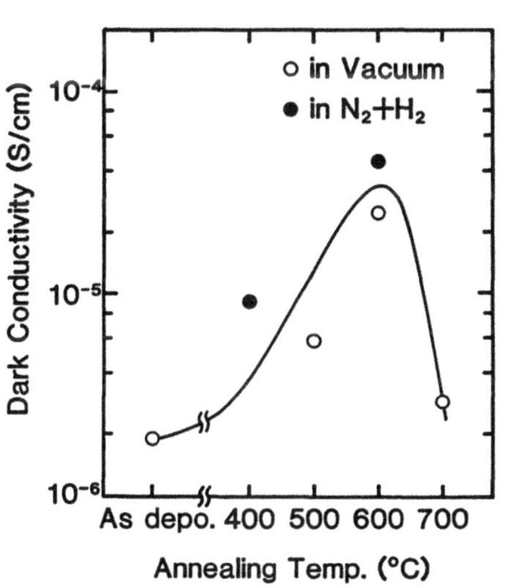

Fig.2 Dark conductivity of a-SiC as a function of annealing temperature

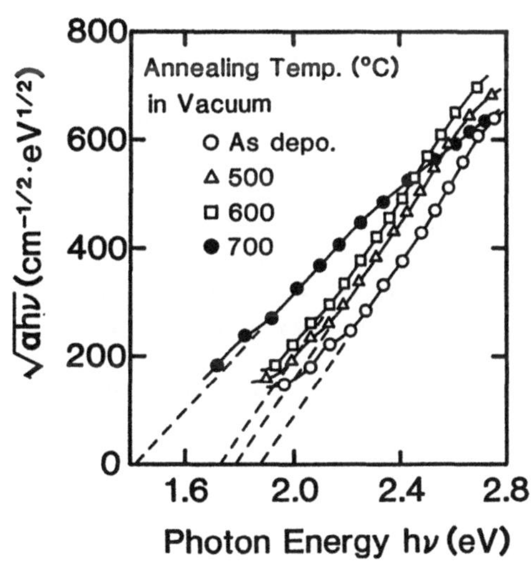

Fig.3 Photo-absorption spectra of a-SiC films annealed at various temperatures

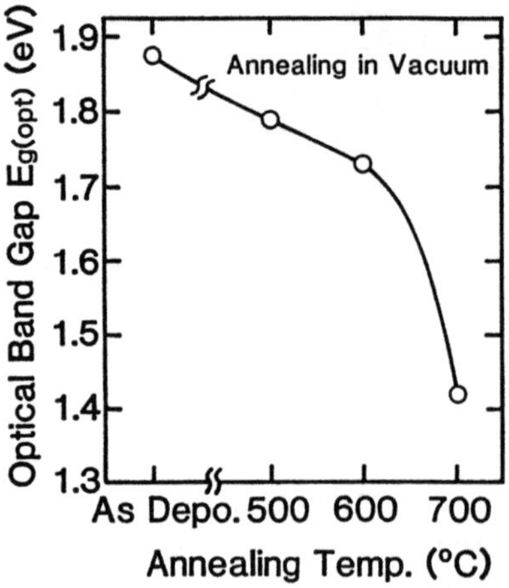

Fig.4 Dependence of optical band gap on annealing temperature

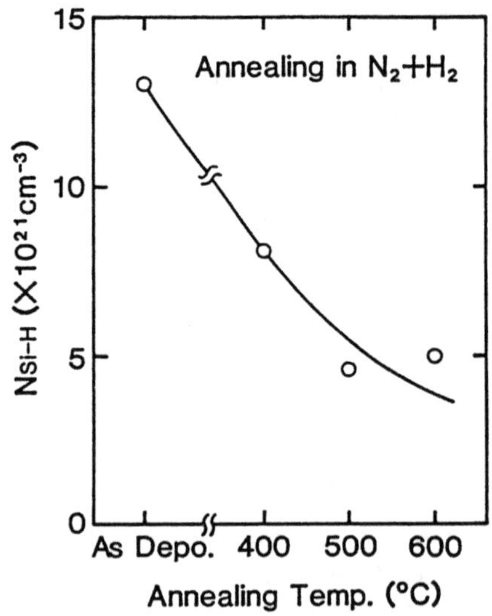

Fig.5 The number of hydrogen atoms bonded to silicon, N_{Si-H}, plotted against annealing temperature

The dissociation of hydrogen atoms was evaluated from IR absorption. Figure 5 demonstrates the number of hydrogen atoms bonded to silicon plotted against annealing temperature. The number of bonded hydrogen atoms was estimated from the absorption area of Si-H wagging mode absorption around $630cm^{-1}$ [4]. It is observed that as-deposited film has fairly high content of bonded hydrogen of $1.3x10^{22}cm^{-3}$ and the bonded hydrogen decreases rapidly with increasing annealing temperature. This suggests that if hydrogen content of as-deposited film can be reduced, the change of physical properties such as conductivities and optical band gaps caused by annealing will be small.

Next, we mention the I-V characteristics of a-SiC/c-Si heterojunction diodes annealed at various temperatures in nitrogen/hydrogen mixed gas ambient. Figure 6a shows the I-V characteristic of the as-deposited diode and Figs.6b-d show those of the diodes annealed at 400-600°C. In order to evaluate the change of the diode characteristics, n-values of the diodes were estimated from the gradient of linear region of log I-V characteristics. Figure 7 shows a relation between n-values of the diodes and annealing temperatures. It is found that n-value improves with increasing annealing temperature. From this result, we can conclude that a-SiC has a thermal stability up to 600°C during device fabrication as for diode characteristics.

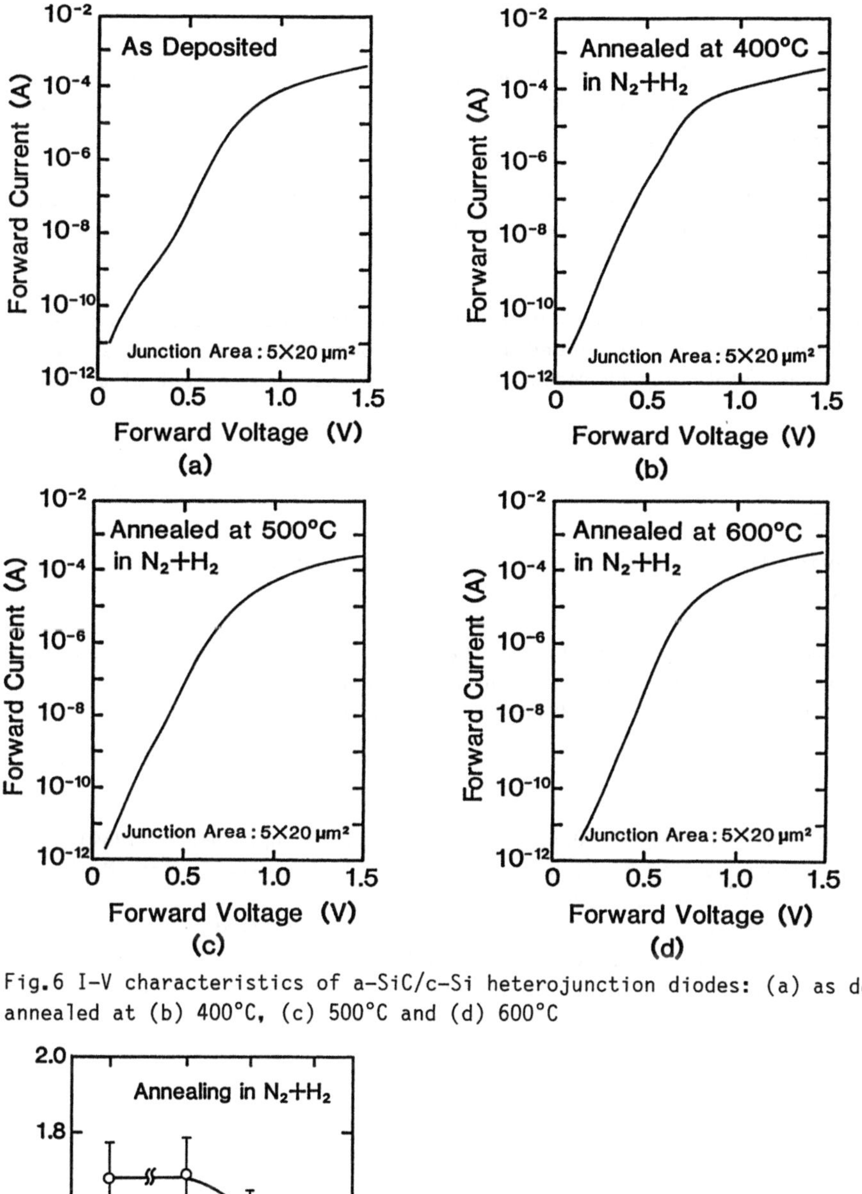

Fig.6 I-V characteristics of a-SiC/c-Si heterojunction diodes: (a) as deposited, annealed at (b) 400°C, (c) 500°C and (d) 600°C

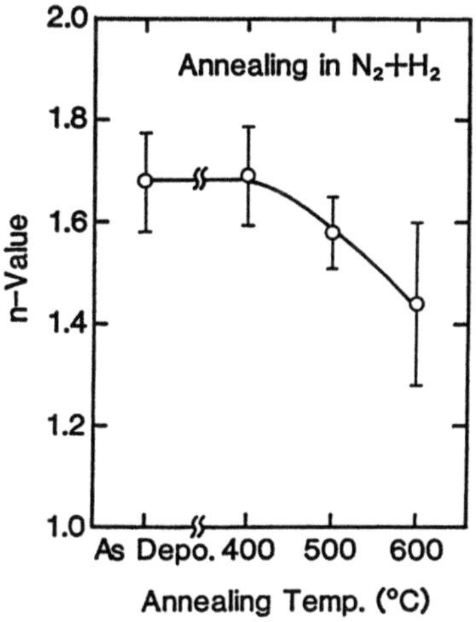

Fig.7 n-value of a-SiC/c-Si heterojunction diode versus annealing temperature

4. SUMMARY

Thermal stability of a-SiC films has been investigated. It is observed that the conductivity increases and the optical band gap slightly decreases with increasing annealing temperature up to 600°C.

Measurements of the I-V characteristics of a-SiC/c-Si heterojunction diodes annealed at various temperatures have shown that n-value improves with increasing annealing temperature up to 600°C.

The change of physical properties of a-SiC caused by heat treatment is considered due to the dissociation of bonded hydrogen atoms and it can be expected to improve the thermal stability by reducing the bonded hydrogen content of as-deposited a-SiC films.

ACKNOWLEDGMENTS

The authors wish to thank Assoc. Prof. H. Ishiwara and Dr. T. Asano for their useful discussions.

This work was partially supported by 1987 Grant-in-Aid for Encouragement of Young Scientist from the Ministry of Education, Science and Culture, NHK Hoso-Bunka Foundation and CASIO Science Promotion Foundation.

REFERENCES

1. M. M. Rahman and S. Furukawa : Jpn. J. Appl. Phys., 23, pp.515-524 (1984)
2. K. Sasaki, S. Furukawa and M. M. Rahman : 1985 Intl. Electron Devices Meeting, Washington D.C., Tech. Dig. pp.294-297
3. K. Sasaki, S. Furukawa and M. M. Rahman : Ext. Abst. 17th Conf. Solid State Devices and Materials, Tokyo, pp.385-388 (1985)
4. C. J. Fang, K. J. Grantz, L. Ley, M. Cardona, F. J. Demond, G. Muller and S. Kalbitzer : J. Non-Cryst. Solids, 35 & 36, pp.255-260 (1980)

Heterojunction Band Discontinuities in Crystalline Silicon/Amorphous Compound Heterojunctions

S.G. Meikle, Y. Hatanaka, Y. Suzuki, and G. Shimaoka

Graduate School of Electronic Science and Technology,
Research Institute of Electronics, Shizuoka University,
3-5-1 Johoku, Hamamatsu 432, Japan

The heterojunction band discontinuities for c-Si/a-$Si_{1-x}N_x$ and c-Si/a-$Si_{1-x}C_x$ heterojunctions have been investigated using the internal photoemission technique for composition and band gap ranges of $0<x<0.45$ and 1.8-2.7eV, respectively, for a-$Si_{1-x}N_x$ and $0<x<0.6$ and 1.8-2.6eV, respectively, for a-$Si_{1-x}C_x$. The conduction band discontinuity is found to remain constant at 0.1eV for nitrogen but increases in the case of carbon. Over the same ranges of x there is a 0.7±0.2eV shift in the X-ray photoelectron spectrum valence band edge position for a-$Si_{1-x}N_x$ and 0.1±0.2eV shift for a-$Si_{1-x}C_x$. Therefore, expansion in the band gap is accommodated by expansion of the valence band discontinuity for c-Si/a-$Si_{1-x}N_x$ and by the conduction band discontinuity for c-Si/a-$Si_{1-x}C_x$.

Introduction:

The band discontinuities that exist in heterojunctions can be used to modify the properties of semiconductor devices. For example, in amorphous silicon solar cells, the large conduction band discontinuity (ΔE_V) in the hydrogenated amorphous silicon/amorphous silicon carbide (a-Si/a-$Si_{1-x}C_x$) heterojunction is used to limit back diffusion of electrons(1).

Because of their practical importance, we have also been studying the band discontinuities in amorphous heterojunctions. Using X-ray photoelectron spectroscopy (XPS) and internal photoemission , the band discontinuities in the crystalline silicon/hydrogenated amorphous silicon nitride (c-Si/a-$Si_{1-x}N_x$) and c-Si/a-$Si_{1-x}C_x$ heterojunctions have been investigated for composition ratios ranging between $0<x<0.45$ and $0.2<x<0.6$ respectively. Our purpose was to determine how the band discontinuities develop as the composition ratio increases from x=0. For the x=0 case (ie.c-Si/a-Si:H) ΔE_C values of 0.1-0.2eV and valence band discontinuity values (ΔE_V) between 0.5-0.7eV have been reported(2,3).

Experimental Methods:

The internal photoemission process has been described elsewhere(4) and will be mentioned only briefly here. As fig. 1 shows, photoexcited carriers with sufficient energy are injected past the interface barrier and measured as a photocurrent I_p. The quantum yield of the photoemission process is given by

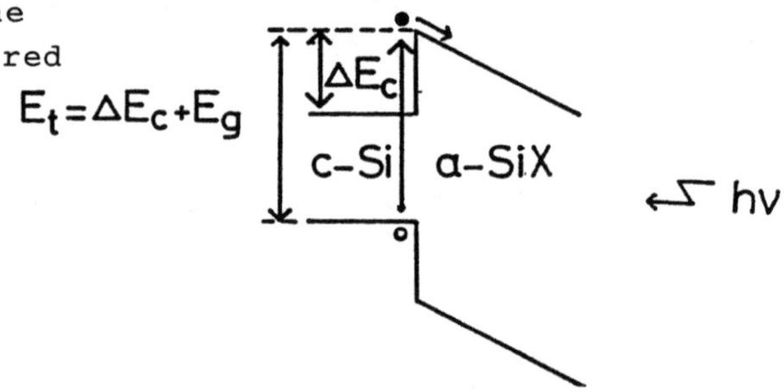

Fig. 1 The internal photoemission process.

$$Y \propto (E_t - h\nu)^a$$

where E_t is the barrier threshold energy, and $h\nu$ is the photon energy. The value for "a" varies between 3/2 and 3 depending on the model being considered(5,6). In our calculations, we used a value of 5/2, proposed by KANE(5), corresponding to an indirect volume excitation process with an imperfect surface. Thus,

$$(I_p/n_{ph})^{2/5} \propto E_t - h\nu$$

where a plot of $(I_p/n_{ph})^{2/5}$ versus $h\nu$ should yield a straight line with an intercept at E_t.

The XPS measurements were performed on an ULVAC PHI-5100 system using the 1253.6eV magnesium K line for the X-ray source. The accuracy in determining band edges was estimated to be ±0.2eV. Intensity data were calculated by integration of the area under the spectral peaks and were corrected for differences in atomic sensitivity(7). For the $a\text{-}Si_{1-x}C_x$ data, background from inelastic scattering has been subtracted.

The films were deposited using a capacitively coupled RF system. For $a\text{-}Si_{1-x}N_x$, the source gases were SiH_4 and 10% NH_3 in N_2 and film composition was controlled by varying the NH_3/SiH_4 ratio. Depositions were performed at 250°C with the pressure and power ranging between 0.04-0.38Torr and 0.5-3.0W, respectively. The $a\text{-}Si_{1-x}C_x$ films were deposited at 0.9Torr and 5W at a temperature of 230°C. The gas mixture consisted of 90% H_2 and 10% C_2H_4+SiH_4 where the C_2H_4/SiH_4 ratio ranged between 1 and 16. Heterojunctions were constructed from 200-300Å amorphous layers deposited on 0.02 ohm-cm (100)p^+ substrates. Contacts consisted of a 150-200Å a-Si:H layer doped using PH_3 and an evaporated semitransparent Au electrode. Prior to deposition, substrates were etched in 10:1 HNO_3:HF to remove native oxide and surface defects.

Experimental Results:

A) c-Si/a-$Si_{1-x}N_x$

Figure 2 shows the internal photoemission data for heterojunctions corresponding to x=0.3 and x=0.45.

The data in fig. 2 indicate that although x rises from 0.3 to 0.45, corresponding to a 0.4eV increase in the band gap, the value for E_t stays constant at about 1.2eV. Table 1 shows that this was also the case for x<0.3. Assuming a silicon band gap of 1.1eV then the value for ΔE_c is approximately 0.1eV.

If ΔE_c is not changing, then the 0.85±0.1eV expansion of the band gap must be accommodated by an increase in ΔE_v. Attempts to confirm this through internal photoemission experiments were not successful because of the difficulty in separating the photoconduction current of the a-$Si_{1-x}N_x$ from the photoemission current. As an alternative, we have monitored the valence band edge position of the a-$Si_{1-x}N_x$ as determined from the XPS spectra. Figure 3 shows these spectra for three different stoichiometries. As x increases from 0 to 0.45, the edge position is seen to shift by 0.7±0.2eV (see insert). This shift, within experimental error, matches the band gap expansion and therefore supports the conclusion that E_v is growing while E_c is

Tab. 1 Summary of composition and photoemission data for a-$Si_{1-x}N_x$.

X	E_g	E_t
0	1.85 eV	1.2 eV
0 2	2.1	"
0 3	2.3	"
045	2.7	"

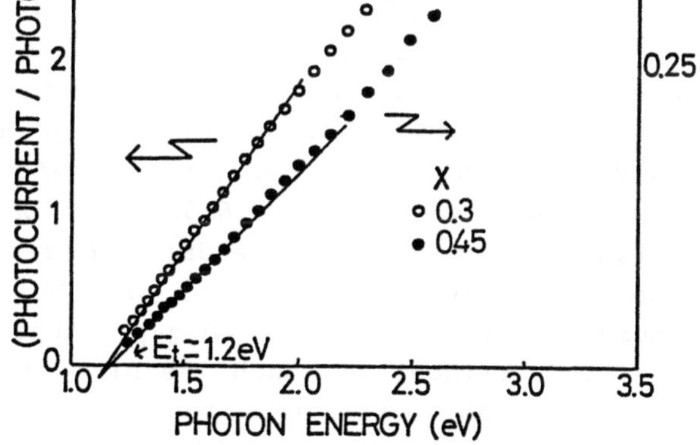

Fig. 2 Internal photoemission data for c-Si/a-$Si_{1-x}N_x$ structures.

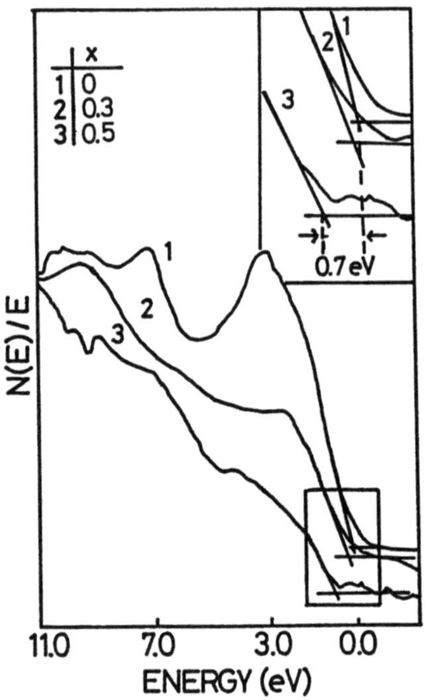

Fig. 3 XPS data from $a\text{-}Si_{1-x}N_x$ for three nitrogen compositions.

constant. Note that our interpretation of the XPS spectra assumes that the $a\text{-}Si_{1-x}N_x$ fermi level does not change. Over the same range of x the N1s peak position shifts toward higher electron energy by approximately 0.2eV. This is expected to be equivalent to the shift in E_f indicating that E_f does change slightly with composition.

B) $c\text{-}Si/a\text{-}Si_{1-x}C_x$

Figure 4 shows an example of photoresponse data taken from a 2.2eV $a\text{-}Si_{1-x}C_x$ sample. For comparison, the 2.3eV $a\text{-}Si_{1-x}N_x$ data of fig. 2 is also included. An intercept in the $a\text{-}Si_{1-x}C_x$ data at 2.2eV is observed which is due to absorption in the thin film. An intercept below the band gap energy, such as that in the $a\text{-}Si_{1-x}N_x$ sample, cannot be seen. Three stoichiometries were investigated corresponding to x=0.8,0.6,0.4 with bandgaps of 2.0eV ,2.2eV, 2.6eV respectively, and all the data were similar to that in fig. 4, having one intercept near the band gap, but no clearly definable E_t at energies lower than the band gap. Apparently a barrier has developed at the interface, although its size cannot be estimated because there is no obvious contribution to the current from internal photoemission.

The results of fig. 4 imply that when ΔE_c becomes large, the probability of photoemission at the interface becomes small. In an attempt to enhance the photoemission component, a large positive bias(0.75V) was applied to the Au electrode. Figure 5 shows the

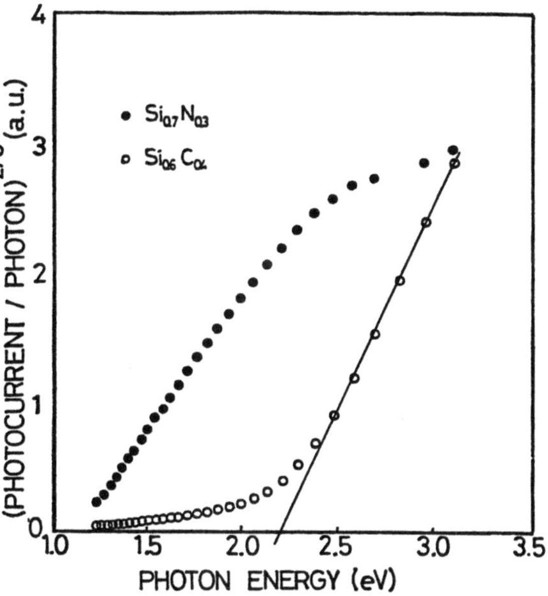

Fig. 4 Comparison of internal photoemission data from a-$Si_{1-x}N_x$ (2.3eV) and a-$Si_{1-x}C_x$(2.2eV) samples.

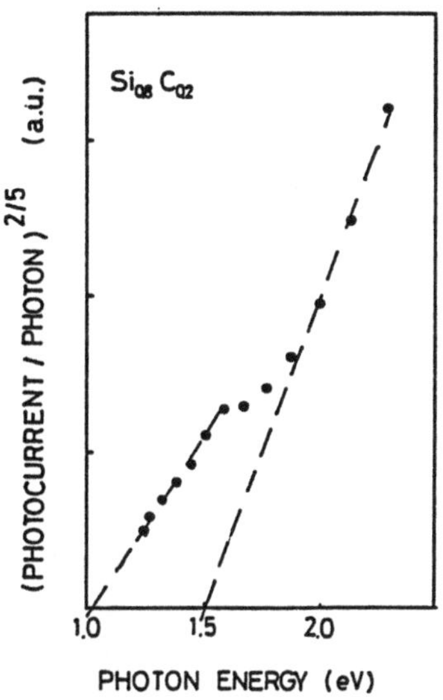

Fig. 5 Photoemission data for the x=0.2(2.0eV) a-$Si_{1-x}C_x$ sample at 0.75V bias.

results for the 2.0eV (x=0.2) sample. From this data, E_t values can be defined at 1.05eV and 1.45eV. The former is due to excitation of carriers from the depletion field in c-Si which has grown as a result of the large bias. If the latter is assumed to be due to photoemission current, then a value of 0.35eV for ΔE_c is obtained. Therefore, ΔE_c has grown while ΔE_v has not increased and may actually have grown smaller.

XPS spectra for the a-$Si_{1-x}C_x$ films are shown in fig. 6. Unlike the case for a-$Si_{1-x}N_x$, these data show very little change in the

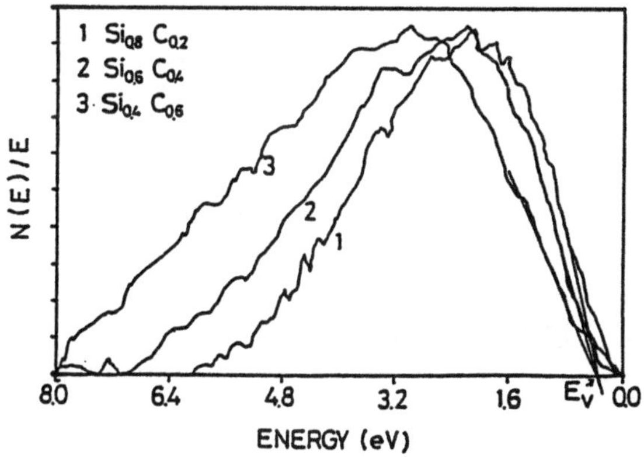

Fig. 6 XPS spectra from a-$Si_{1-x}C_x$ showing valence band edge position.

valence band edge position (<0.1eV), implying that for the case of carbon the expansion of the band gap is accommodated by movement in the conduction band.

Discussion and Summary:

The results for the c-Si/a-$Si_{1-x}C_x$ junction are similar to those reported for the a-Si/a-$Si_{1-x}C_x$ junction (8,9). In all cases the results indicate a large ΔE_c. However, the a-Si/a-$Si_{1-x}N_x$ junction apparently differs from the c-Si/a-$Si_{1-x}N_x$ case investigated here. COLUZZA et al. report values for ΔE_c and ΔE_v of 0.5eV and 0eV respectively(10) for the all-amorphous junction. On the other hand, in agreement with what we observe, KAERCHER et al. report, from XPS studies of a-$Si_{1-x}N_x$, almost no shift in the conduction band position over a range corresponding to 0-50% nitrogen(11).

In conclusion, we find that for the crystalline silicon/amorphous compound heterojunction the behavior of the heterojunction discontinuities for a-$Si_{1-x}C_x$ is opposite to a-$Si_{1-x}N_x$. For a-$Si_{1-x}N_x$ corresponding to $0<x<0.45$, ΔE_c remains constant at approximately 0.1eV while ΔE_v grows to accommodate the expanding band gap. For a-$Si_{1-x}C_x$ ΔE_c expands and ΔE_v remains relatively constant.

References:

1. Y. Tawada, H. Okamoto, and Y. Hamakawa, Appl Phys. Lett. 39, 237 (1981).
2. H.Mimura and Y.Hatanaka,Appl.Phys.Lett. 50, 326 (1987).
3. H. Matsuura, T. Okuno, H. Okushi, and K. Tanaka, J. Appl. Phys. 55, 1012 (1984).
4. S.G. Meikle, Y. Hatanaka, Y. Suzuki, and G. Shimaoka : Proc. of the fourth International Conference on Solid Films and Surfaces, J. Appl. Surf. Sci., to be published.
5. E.O. Kane, Phys Rev. 127, 131 (1962).
6. R.J. Powell, J. Appl. Phys. 41, 2424 (1970).
7. Handbook of X-ray Photoelectron Spectroscopy, C.D. Wagner et al.: Perkin-Elmer Corporation Physical Electronics Division.
8. F. Evangelisti, J. Non-Cryst. Sol., 77&78, 969 (1985), references therein.
9. Y. Okayasu, K. Fukui, and M. Matsumura, Appl. Phys. Lett., 50, 248, (1986).
10. C. Coluzza, G. Fortunato, C. Quaresima, M. Capozi, and P. Perfetti, J. Non-Cryst. Sol., 77&78, 999 (1985).
11. R. Kaercher, L. Ley, and R.L. Johnson, Phys. Rev. B 30, 1896, (1984).

Amorphous Silicon Solar Cells Using a-SiC Materials

Y. Kuwano and S. Tsuda

Research Center, Sanyo Electric Co., Ltd.,
1-18-13, Hashiridani, Hirakata, Osaka, Japan

Research improvements in amorphous silicon (a-Si) solar cells using a-SiC are reviewed. After Hamakawa group reported on the valency control of a-SiC and its application to a-Si solar cells, there have been various improvements in the field of a-SiC, such as hydrogen dilution effect, a-SiC:H:F, a-SiC/a-Si superlattice structure and microcrystalline SiC with ECR-CVD method.

Recently, a new type of doping gas, $B(CH_3)_3$, was proposed by the authors for the p-type doping gas of a-SiC, instead of B_2H_6. In the case of a-SiC prepared by photo-CVD with $B(CH_3)_3$, low defect density and high photoluminescence intensity were observed. For GD (glow discharge) with $B(CH_3)_3$, bandgap narrowing, which is a large problem in the case of B_2H_6, is not observed. An intrinsic (effective area) efficiency of 10 % was obtained for a 100 cm^2 a-Si solar cell whose p-layer was a-SiC doped with $B(CH_3)_3$.

a-SiC/a-Si superlattice structure films were deposited with a photo-CVD method by the authors for the first time. In addition to improvements in film properties, the p-type a-SiC/a-Si superlattice structure film was found to be an effective layer for photovoltaic energy conversion.

1. Introduction

Since Spear and LeComber showed valency control of hydrogenated amorphous silicon (a-Si:H) deposited by glow discharge (GD) in SiH_4,[1] extensive research on a-Si and a-Si devices has been performed. In the field of solar cells, Carlson and Wronski reported the first a-Si solar cell in 1976.[2] Since then the a-Si solar cells have been attracting much attention as a low-cost solar cell. In 1981, a p-type a-SiC/in a-Si solar cell was proposed by the Hamakawa group, resulting in a considerable increase of conversion efficiency.[3] The conversion efficiency of the a-Si solar cell was improved to 11.7 % with a high-quality i-layer prepared by a super chamber method and a highly conductive p-layer prepared by a photo-CVD method.[4] To apply an a-Si solar cell as an electric power source, further improvement in the conversion efficiency is required.

In this paper, first, a brief review of a-SiC and related devices is presented. Then, as recent topics, a new doping gas of $B(CH_3)_3$ and a-SiC/a-Si superlattice structure are reported with solar cells. Finally, future prospects in the field of a-SiC are mentioned.

2. Brief review of a-SiC and related devices

In 1977, Anderson and Spear reported a-SiC:H films deposited by a GD method, and showed that the optical bandgap of undoped a-SiC:H increases continuously with increasing carbon content.[5] Valency control of a-SiC was reported by the Hamakawa group in 1981, as well application of p-type a-SiC to window layers of heterojunction in a-Si solar cells.[3] Since then, a number of investigations have been made on a-SiC and related devices, as shown in Table I.

There have been various improvements in preparation methods, doping, and superlattices of a-SiC. In order to avoid the intermixing of impurities, a separated reaction chamber system was adopted for the fabrication of p-a-SiC/in a-Si solar cells by the authors in 1982.[8] In 1984, photo-CVD (Chemical Vapor Deposition) method was applied to the fabrication of the p-a-SiC/in a-Si solar cells by Konagai et al., which eliminated plasma damage caused by GD.[9] In 1985, HR (hydrogen

Table I History of a-SiC and related devices

		Year (1977, 79, 81, 83, 85, 87)
FILMS	Preparation methods	undoped(Dundee); photo-CVD(TIT); ECR(Osaka); valency control(Osaka); H-dilution(ETL); separated chamber(Sanyo); HR-CVD(TIT)
FILMS	Doping	B_2H_6, PH_3(Osaka); $B(CH_3)_3$(Sanyo)
FILMS	Superlattice	undoped, GD(TIT); doped, photo-CVD(Sanyo)
DEVICES	Solar cells	hetero junction(Osaka); superlattice structure(Sanyo); buffer layer (TIT); i-layer (Fuji, Solarex)
DEVICES	Others	EL(TIT); HBT(TIT)

radical) - CVD was proposed for the preparation of hydrogenated-and-fluorinated a-SiC by Shimizu et al.[10] In order to improve film quality of a-SiC prepared by GD, hydrogen - dilution method was applied by Matsuda et al.[11] Recently, the Hamakawa group prepared a-SiC with ECR (Electron Cyclotron Resonance) -CVD, and used it for p-a-SiC/in a-Si solar cells, resulting in an improved conversion efficiency of 11.8 %.[12]

For doping, B_2H_6 was conventionally used to prepare p-type a-SiC.[3] Recently, we proposed a new doping gas, $B(CH_3)_3$, which will be described in the next section.

a-Si superlattice structures, which were reported by Tiedje and Abeles in 1982,[13] have gathered much attention as a new material. In 1983, Kukimoto et al. prepared a-SiC/a-Si superlattice structure films by GD.[14] In order to improve interface properties, a-SiC/a-Si superlattice structure films were prepared by photo-CVD method by the authors in 1985,[15] which will be mentioned in Section 4. In addition to these applications to window layers of a-Si solar cells, a-SiC buffer layers (p/i interface layers),[16] and a-SiC i-layers (bulk layers)[17,18] were also studied as new applications of a-SiC in a-Si solar cells.

Two a-SiC related devices, heterojunction a-Si solar cells and superlattice structure p-layer a-Si solar cells were reported in 1981[3] and 1985,[15] respectively. Other devices, such as EL (electro-luminescence) cells[19], hetero bipolar transistors,[20] and photoreceptors were also investigated.

3. Doping with $B(CH_3)_3$

At present, the p-type a-SiC films which are used in the conventional a-Si solar cells at present have some problems. Namely, the photoconductivity of p-type a-SiC films has been shown to decrease as the optical bandgap increases, whereas both high photoconductivity and wide optical bandgap are required for a high-efficiency solar cell. The conventional p-type doping gas of B_2H_6 has a B-B bond, as shown in Fig. 1. This B-B bond seems to remain easily in a-SiC films, resulting in high defect density. Therefore, for the first time, a new doping gas of $B(CH_3)_3$ is studied.

In this section, a comparison of a-SiC film properties doped with B_2H_6 and $B(CH_3)_3$ is given. The application of high-quality p-type a-SiC to window layers of high-efficiency a-Si solar cells is studied.

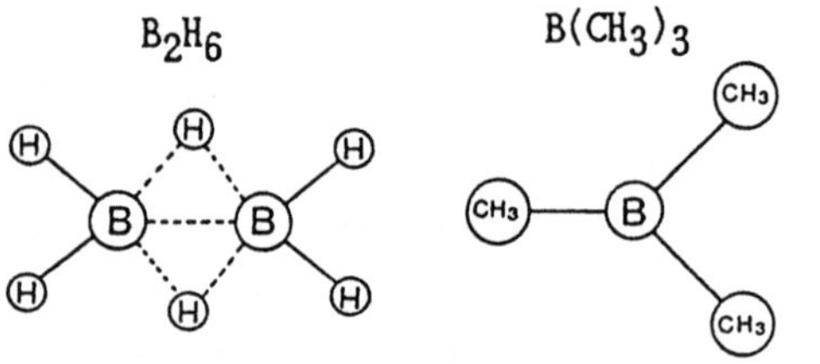

Fig. 1 Molecular structure of B_2H_6 and $B(CH_3)_3$

3-1. Preparation

p-type a-SiC films with $B(CH_3)_3$ are prepared by both a photo-CVD and GD. Typical reaction conditions are shown in Table II. In the photo-CVD method, a low pressure mercury lamp was used as the light source for direct gas decomposition. Si_2H_6 and C_2H_2 gases were used as Si and C sources, respectively. In the GD method, SiH_4 and CH_4 gases were used as Si and C sources, respectively. These pairs of gases are conventional for both method.[3,9] For comparison, p-type a-SiC films with B_2H_6 were also prepared.

Table II Typical reaction conditions of a-SiC films

	Photo-CVD	Glow discharge
Material gas	Si_2H_6 C_2H_2 $B(CH_3)_3$ or B_2H_6	SiH_4 CH_4 $B(CH_3)_3$ or B_2H_6
Doping ratio	0.1~3%	0.1~3%
Substrate temperature	150°C~300°C	150°C~300°C
Deposition	Direct photolysis	C-coupled parallel electrode
Power	184.9nm 5mW/cm^2 253.7nm 30mW/cm^2	13.56MHz 20W

3-2. Film properties of a-SiC prepared by photo-CVD

Figure 2 shows photoluminescence spectra. In the case of B_2H_6, photoluminescence intensity is very small, as explained by creation of non-radiative recombination centers. In the case of $B(CH_3)_3$, however, the photoluminescence intensity is very large; it is almost the same as that of undoped a-SiC. Optical absorption spectra of the films were measured by photothermal deflection spectroscopy. In low energy region corresponding to tail states, the absorption coefficient of a-SiC films prepared with $B(CH_3)_3$ is one order smaller than that of a-SiC films prepared with B_2H_6.[21]

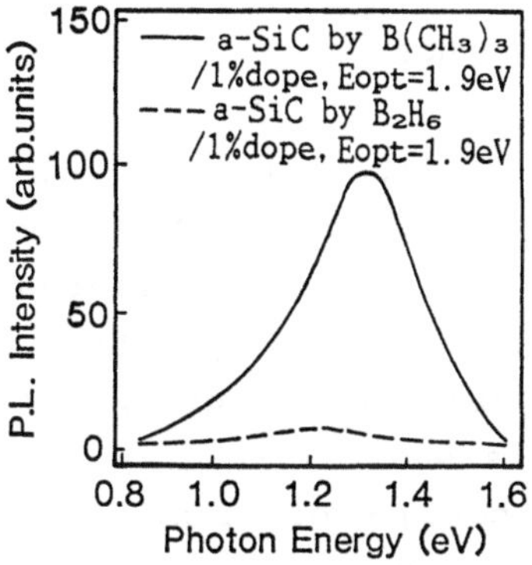

Fig. 2 Photoluminescence (PL) spectra of p-type a-SiC films

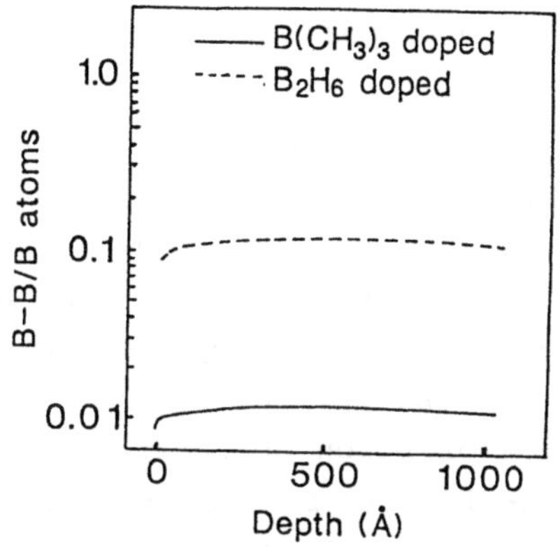

Fig. 3 Concentration ratio of B-B bonds/B atoms of p-type a-SiC films measured by SIMS

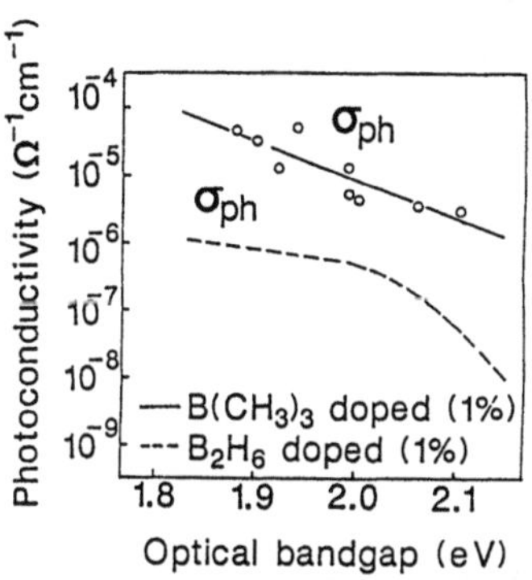

Fig. 4 Photoconductivity of p-type a-SiC films as a function of optical bandgap

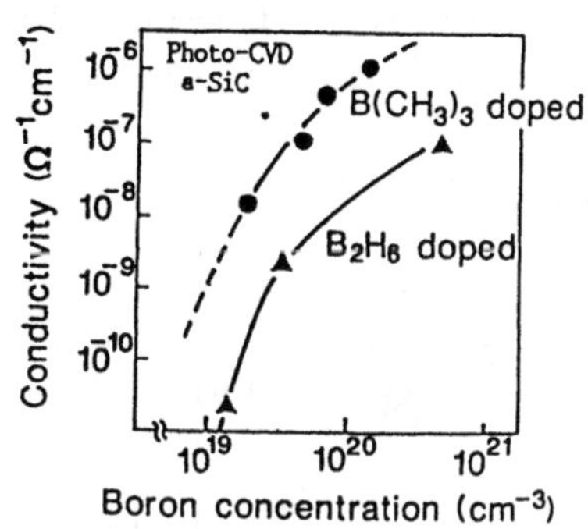

Fig. 5 Dark conductivity of p-type a-SiC films as a function of boron concentration measured by SIMS

To analyze the reason for the improvement in film properties, ratio of B-B bonds/B atoms was measured by SIMS. The ratio of B-B bonds/B atoms in a-SiC prepared with $B(CH_3)_3$ is about ten times smaller than that with B_2H_6, as shown in Fig. 3. It was found that the increase in the ratio of B-B bonds/B atoms relates to the increase in defect density.

The photoconductivities of a-SiC films are improved one or two orders of magnitude by using $B(CH_3)_3$ instead of B_2H_6, as shown in Fig. 4. The dark conductivities of a-SiC films (E_{opt} = 2.0 eV) as a function of boron concentration are shown in Fig. 5. At the same boron concentration, the dark conductivity of a-SiC films with $B(CH_3)_3$ is two orders higher than that with B_2H_6. Therefore excellent boron doping efficiency is obtained by using a new doping gas of $B(CH_3)_3$. It is considered that the higher photoconductivity and the higher doping efficiency of a-SiC with $B(CH_3)_3$ are related to the lower B-B bond density.

3-3. Film properties of a-SiC prepared by GD method

To confirm the superiority of $B(CH_3)_3$ gas, a-SiC films fabricated by the glow discharge method have also been studied. The same improvements in a-SiC film qualities are expected. The conventional a-SiC films fabricated by the glow discharge method with B_2H_6 have a problem of the bandgap narrowing at the highly doped region. The change in the optical bandgap of a-SiC films is shown in Fig. 6 as a function of boron concentration. The optical bandgap of highly boron doped a-SiC with B_2H_6 decreases as the boron concentration increases.

In constrast, the optical bandgap of a-SiC films with $B(CH_3)_3$ is almost constant ($\sim$ 2.0 eV). Namely, the bandgap narrowing can be avoided by using $B(CH_3)_3$. At the same optical bandgap of 2.0 eV, the photoconductivity of a-Si film with $B(CH_3)_3$ is also improved from $1 \times 10^{-6}\ \Omega^{-1}\ cm^{-1}$ to $2 \times 10^{-5}\ \Omega^{-1}\ cm^{-1}$ compared to that with B_2H_6.

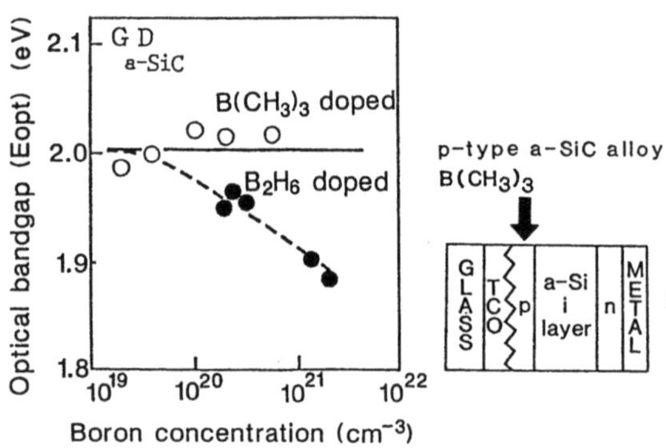

Fig. 6 Optical bandgap of p-type a-SiC films fabricated by GD method as a function of boron concentration

3-4. Solar cell characteristics

As mentioned above, the use of $B(CH_3)_3$ instead of B_2H_6 is effective for p-type a-SiC films and high-quality a-SiC films are obtained with $B(CH_3)_3$. So pin a-Si solar cells whose p-layers were doped with $B(CH_3)_3$ were studied.

Fig. 7 shows illuminated I-V characteristics of an integrated type a-Si submodule (10 cm x 10 cm), whose p-layer was fabricated by the GD method with $B(CH_3)_3$. A conversion efficiency of 10.0 % (practical (total area) efficiency of 9.02 %) has been obtained. A high-quality i-layer by the super chamber and laser patterning method are also effective for achieving a high conversion efficiency.

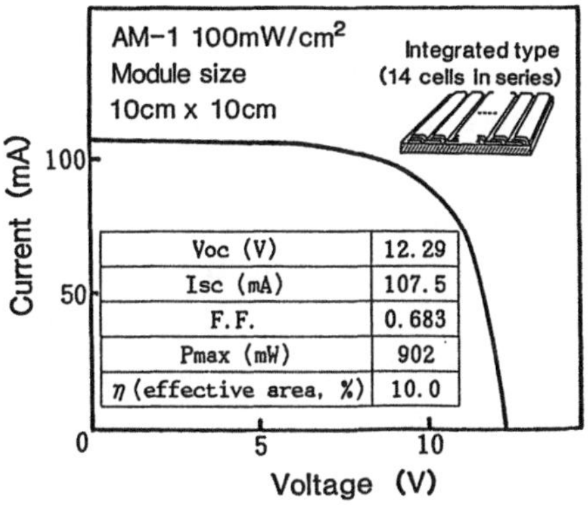

Fig. 7 Illuminated I-V characteristics of an integrated type a-Si solar cell submodule

4. a-SiC/a-Si superlattice structures

Previous amorphous superlattice structure films have been fabricated by the GD method.[13,14] It is thought that, when this method is used, interface properties are deteriorated by high energy particles in the plasma.[9] Therefore, a-Si superlattice structure films were fabricated by the photo-CVD method,[15] and a new type of solar cell was developed using the superlattice structure as the p-layer of an a-Si solar cell.

4-1. Film properties

Samples are periodically alternated between a-Si:H (well layers) and a-SiC:H (barrier layers).[22] In order to analyze the structure of the superlattice, a-Si:H of 25 Å and a-SiC:H of 25 Å were deposited by the photo-CVD method. The depth profiles of composition atoms (Si, C) were measured by AES (Auger electron spectroscopy), as shown in Fig. 8. The atomic concentration of Si and C varies periodically, which corresponds to our design. In X-ray diffraction of the films, the Bragg reflection peak also corresponds to designed period.

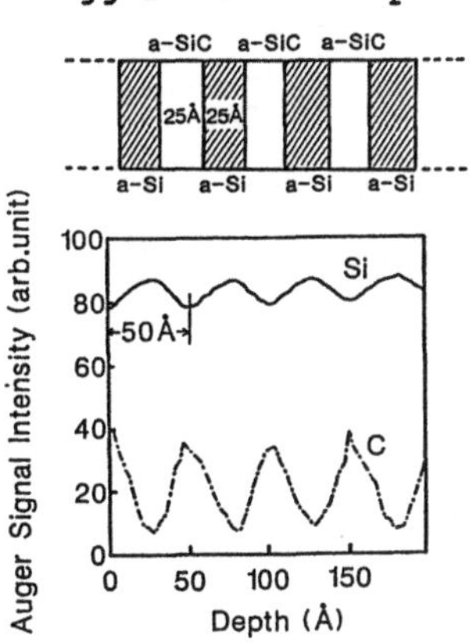

Fig. 8 Depth profiles of Si and C in an a-Si/a-SiC superlattice structure film measured by AES (Auger Electron Spectroscopy).

In order to investigate the effect of plasma damage on interfaces in a-Si films, photoluminescence spectra were measured for amorphous silicon carbide alloy, a superlattice structure film deposited by the GD method, and a superlattice structure film deposited by the photo-CVD method. The intensity of the superlattice structure film is about 100 times larger than that of the alloy, as shown in Fig. 9. In addition, the intensity of the superlattice structure film by the photo-CVD method is about 4 times larger than that by the GD method. These results show that the photo-CVD method causes less damage to the interface than the GD method. Good superlattice structure film was thus obtained by using the photo-CVD method.

Figure 10 shows the optical bandgap (E_{opt}) and B-value as a function of the well layer thickness. E_{opt} and B-value were determined by the Tauc plot. The barrier layer thickness was fixed at 25 Å, and the well layer thickness was varied from 8 Å to 500 Å. As the well layer thickness was decreased below 50 Å, E_{opt} increased from 2.0 eV to 2.2 eV and B-value increased from 900 ($eV^{1/2}$ $cm^{-1/2}$) to 1100 ($eV^{1/2}$ $cm^{-1/2}$). These blue shifts were thought to be caused by the separation of the lowest quantum level.

The photoconductivity and dark conductivity vertical to the interfaces of the a-Si/a-SiC superlattice structure films were measured. In the wide optical bandgap region, the photoconductivity of the superlattice structure films was much higher than that of a-SiC:H films prepared by the GD method.

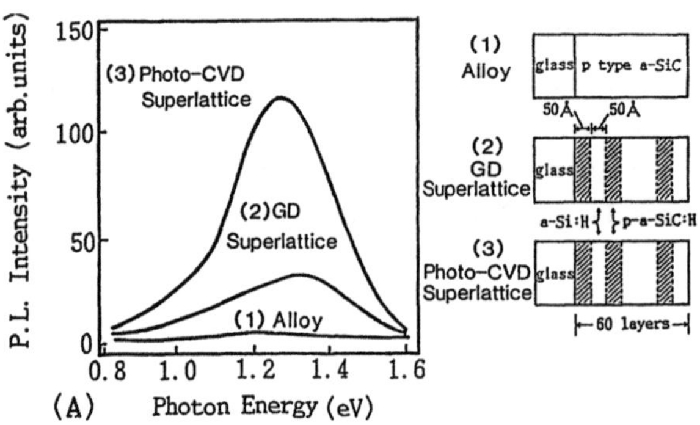

Fig. 9 Cross sections and the photoluminescence spectra of superlattice films and an alloy film.

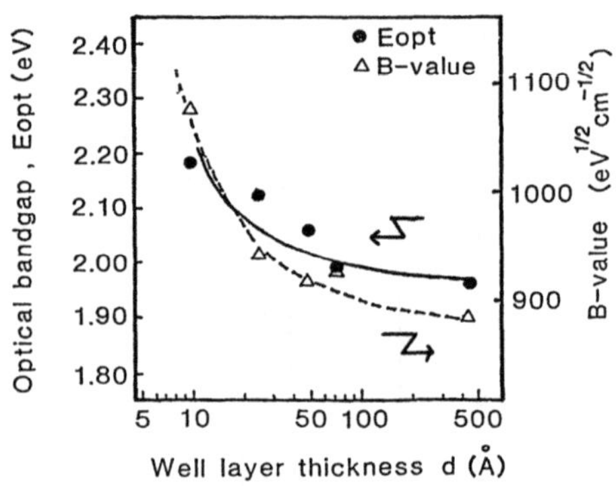

Fig. 10 E_{opt} and B-value of a-Si/a-SiC superlattice structure films fabricated by the photo-CVD method as a function of the layer thickness.

4-2. Application to solar cells

We have also developed a new type of cell using the superlattice structure film as the p-layer of pin a-Si solar cells. In this cell, short circuit current is greatly increased for an increase in the collection efficiency in short wavelength region, as indicated by a hatched area in Fig. 11.

The increase in the collection efficiency can be explained by assuming that the superlattice structure p-layer is an effective area for photovoltaic effect. In order to ascertain this effect, we fabricated pn junction solar cells, which do not have i-layers as shown in Fig. 12, and investigated their characteristics. When the p-layer of superlattice structure film prepared by the photo-CVD method was used, the inner quantum efficiency was five times greater than that of a coventional a-SiC p-layer prepared by the GD method, as shown in Fig. 12.

Superlattice structure p/in a-Si solar cells were fabricated using $B(CH_3)_3$ to improve film quality, A conversion efficiency of 11.2 % was obtained for a 1 cm^2 cell, and it will be improved by optimizing fabrication conditions.

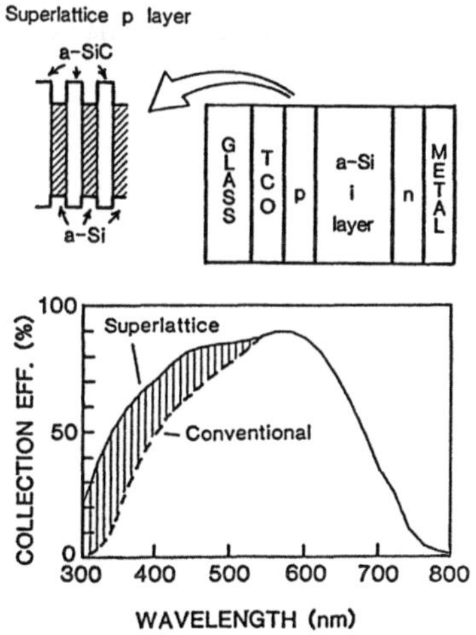

Fig. 11 Collection efficiency of the superlattice structure p-layer a-Si solar cell

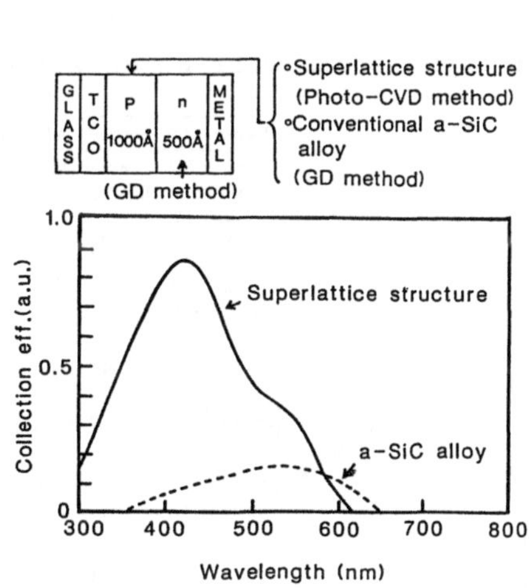

Fig. 12 Collection efficiency spectra of p-n junction cells

5. Future prospects

a-SiC is now regarded as a promising material for future electronic devices. Here, future trends in a-SiC and related devices are discussed with Fig. 13.

First, further improvements in quality of a-SiC films are required. For this purpose, research on preparation method for advanced CVD will be made, and starting materials, such as higher silane, fluorinated silane, and new carbon sources will be considered. Special emphasis will be placed on purification, because high-quality material is sensitive for impurities. In the case of a-Si, super chamber (separated ultra-high vacuum reaction chamber) method was proposed by the authors,[23] and it will be adopted in preparation of high quality a-SiC.

Doping gases and structures, which were mentioned before, will also be improved. As for structure, reduction of microscopic defects which are thought to have the nature of alloys is important.

With high-quality a-SiC films, high-performance a-Si(C) devices will be developed. In solar cells, they will be used not only in the window layer, but in buffer layer at p/i interface and bulk layer for multi-bandgap cells. In other devices, ultraviolet photosensors,[24] photoreceptors with SiC surface layer, and transistors such as hetero bipolar transistors will be developed.

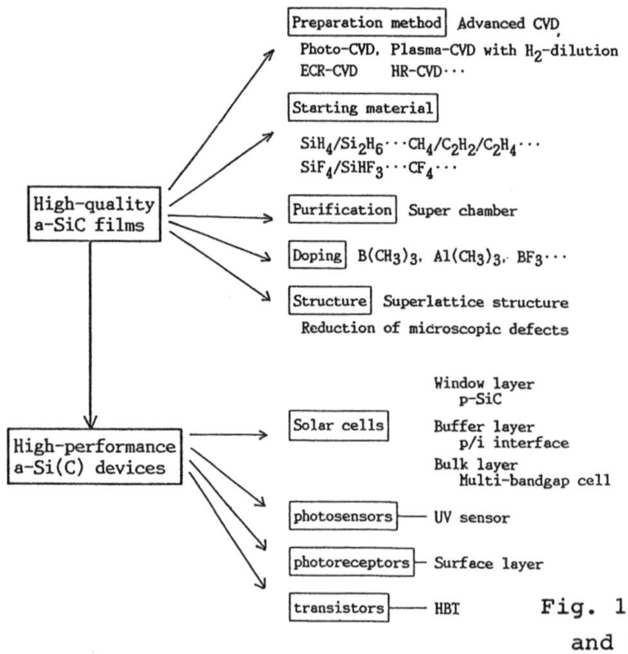

Fig. 13 Future trends in a-SiC and related devices

6. Conclusions

Research improvements in a-Si solar cells using a-SiC were reviewed. As recent topics, a new doping gas of $B(CH_3)_3$ was proposed instead of B_2H_6, and a-SiC/a-Si superlattice structure was studied to improve properties of wide-bandgap material. Future trends in a-SiC and related devices were also mentioned.

Acknowledgement

This work is supported in part by New Energy Development Organization as a part of the Sunshine Project under the Ministry of International Trade and Industry.

Reference

1) W. E. Spear and P. G. LeComber : Solid State Commun. 17 (1975) 1198

2) D. E. Carlson and C. R. Wronski : Appl. Phys. Lett. 28 (1976) 671

3) Y. Tawada, Y. Hamakawa et al. : Proc. 15th IEEE Photovoltaic Specialists Conf., Florida (1981) p.245

4) S. Nakano et al. : Proc. Materials Research Society, Palo Alto (1986) p.511

5) D. A. Anderson and W. E. Spear : Phil. Mag. 35 (1977) 1

6) Y. Kuwano et al. : Proc. 15th IEEE Photovoltaic Specialists Conf., Florida (1981) p.698

7) Y. Kuwano et al. : Jpn. J. Appl. Phys. 21 (1982) 413

8) Y. Kuwano et al. : Proc. 4th E. C. Photovoltaic Solar Energy Conf., Stresa (1982) p.704

9) T. Tanaka, M. Konagai et al. : Technical Digest of Int'l PVSEC-1, Kobe (1984) p.563

10) S. Oda, I. Shimizu et al. : J. Non-Crystalline Solids 77 & 78 (1985) 877

11) A. Matsuda et al. : Jpn. J. Appl. Phys. 60 (1986) 4025

12) Y. Hattori, Y. Hamakawa et al. : Proc. 19th IEEE Photovoltaic Specialists Conf., New Orleans (1987) p.689

13) B. Abeles and T. Tiedje : Phys. Rev. Lett. 51 (1983) 2003

14) H. Munekata and H. Kukimoto : Jpn. J. Appl. Phys. 22 (1983) L544

15) S. Tsuda et al. : Proc. 18th IEEE Photovoltaic Specialist Conf., Las Vegas (1985) p.1295

16) W. Y. Kim et al. : Extended Abstract of 17th Conf. Solid State Devices and Materials, Tokyo (1985) p.4

17) A. Catalano et al. : Proc. 19th IEEE Photovoltaic Specialists Conf., New Orleans (1987) p.1095

18) T. Yoshida et al. : Proc. 19th IEEE Photovoltaic Specialists Conf., New Orleans (1987) p.1095

19) H. Munekata and H. Kukimoto : Appl. Phys. Lett. 42 (1983) 432

20) K. Sasaki, S. Furukawa et al. : Electron Devices Lett. EDL-6(6) (1985) 311

21) H. Tarui et al. : Technical Digest of Int'l PVSEC-3, Tokyo (1987) p.41

22) S. Tsuda et al. : Jpn. Appl. Phys. 26 (1987) 28

23) S. Tsuda et al. : Jpn. Appl. Phys. 26 (1987) 33

24) S. Nakano et al. : Proc. 4th Int'l Conf. Solid-State Sensors and Actuators, Tokyo (1987) p.245

Au-Ni Contacts on ß-SiC Films

V.H. Watt and K.H. Jackson

Materials Science Research Center of Excellence, Howard University, Washington, DC 20059, USA

In order for SiC to realize its potential as an electronic material the necessary contact technology must be developed. Auger Electron Spectroscopy (AES), and the Cox and Strack contact resistance measurement techniques were used to study Ni-Au contacts on ß-SiC. We observed that contacts formed by depositing 1000Å Ni and 2000Å Au display rectifying qualities. They become ohmic with contact resistance, r_c, between 2 - 6 X 10^{-3} ohm/cm^2 when annealed between 300°C and 700°C. AES results indicate the build-up of Ni which appears to block Au interdiffusion at temperatures less than 400°C, while considerable diffusion of Au into the epitaxial layer was observed at temperatures greater than 500°C. The Ni acts as a diffusion barrier and the Au forms the ohmic contact. When the contact is annealed, the Ni forms a silicide, which is about 10 - 30Å deep. What happens to the displaced C atoms and their effect on the electrical properties of the material will be discussed in this paper.

Springer Proceedings in Physics, Vol. 34 **Amorphous and Crystalline Silicon Carbide**
Editors: G.L. Harris and C.Y.-W. Yang

Epitaxial Growth, High Temperature Ion Implantation and MOSFET Fabrication in Monocrystalline ß-SiC Thin Films

H.S. Kong, J.A. Edmond, J.W. Palmour, J.T. Glass[1], and R.F. Davis

Department of Materials Science and Engineering, Box 7907,
North Carolina State University, Raleigh, NC 27695-7907, USA

1. ABSTRACT

High purity β-SiC films have been epitaxially grown on Si (100) and α-SiC (0001) at 1633K–1823K and 0.1 MPa using SiH_4 and C_2H_4 carried in H_2. Films grown on the on-axis Si (100) contained substantial concentrations of dislocations, stacking faults and antiphase boundaries; those deposited on α-SiC (0001) contained primarily double positioning boundaries. These boundaries were eliminated by using off-axis orientations of the respective substrates. Films produced on Si (100) have also been doped via high temperature ion implantation with Al (p-type) and N(n-type). Transmission electron microscopy analysis showed a markedly reduced defect and precipitate concentration relative to that observed in similar research at the lower temperatures. MOSFET devices have been fabricated and their current-voltage characteristics measured from 298 to 923K. Tansconductances as high as 11.90 mS/mm were achieved.

2. INTRODUCTION

Monocrystalline Si has been almost universally adopted as the substrate of choice for the growth of the β-SiC thin films. Amelioration of the ~8% and 20% mismatches in the coefficients of thermal expansion and lattice parameters, respectively, via the initial reaction of the Si (100) surface with a C-containing gas to form a thin layer of SiC [2] followed by epitaxial deposition of the β-SiC films using individual C- and Si-containing gases has been reported by several investigators [3–7]. For a description of the experimental steps and/or the analysis of the results of these procedures, the reader is referred to Ref. [2] and to selected papers in these Proceedings in which growth is a major issue. Beta-SiC films have also been previously grown on monocrystalline α-SiC substrates in the temperature range of 1773K–1973K [8,9]. However, the growth conditions were not optimized and the interface between the SiC epilayer and the SiC substrates was not investigated.

Edmond et al. [10] have performed room temperature implants in β-SiC followed typically by a 2073K–300s anneal. This latter procedure resulted not only in electrical activation and recrystallization of the lattice but also the formation of residual defects. The percentages of ionized and activated dopants of ≈20%, 0.5% and 35% for P, Al and N, respectively, were ascertained from electrical measurements conducted on samples at room temperature. These results were not improved by implanting at liquid N_2 temperature. Thus high temperature studies were initiated to achieve simultaneously structural annealling and activation of the dopant atoms.

Shibahara et al. [11] and Kondo et al. [12] have produced working metal oxide semiconductor transistors (MOSFETs) on β-SiC films. Their devices functioned in either the enhancement or the depletion mode. However, considerable leakage current through the p-type layer was evident, as the drain current did not saturate even at drain voltages >-12 V.

The following sections detail the procedures and results of (i) growth of β-SiC on Si (100) and α-SiC (0001), including the observation of defect structures in the grown films, (ii) high temperature ion implantation, and (iii) MOSFETs fabrication and characterization.

3. EXPERIMENTAL PROCEDURES

The considerable mismatch in the lattice parameters of Si (100) and β-SiC (100) noted above resulted in the formation of numerous microtwins, intrinsic stacking faults and antiphase boundaries

Springer Proceedings in Physics, Vol. 34 **Amorphous and Crystalline Silicon Carbide**
Editors: G.L. Harris and C.Y.-W. Yang

(APB). Shibahara et al. [13] have recently shown that APBs can be successfully eliminated in β-SiC using off-axis Si (100) with the [100] inclined 2° toward [011]. In our research both 2° and 4° inclinations of the Si [100] toward [011] were investigated. Unfortunately, the other types of defects were not eliminated by this procedure. Therefore, in an attempt to eliminate all of the various defects simultaneously, growth directly on and 3° off the Si (0001) and C (000$\bar{1}$) faces of Acheson-derived 6H α-SiC substrates has been studied within the temperature range 1683K–1823K at 1 atm total pressure and using a SiH_4/C_2H_4 flow ratio of 2. For a complete description of experimental procedures for both of these approaches the reader is referred to Refs. [14,15].

The dopants of Al and N were individually incorporated into the films via ion implantation at elevated temperatures using an offset angle of 7°. Substantial *in situ* annealing occurred during high temperature implantation; thus a much lower post implantation annealing temperature of 1473K (for 1800 s) was sufficient to maximize the activation of the dopants. For a complete description of the experimental high temperature implant procedures and conditions and the methods of characterization, the reader is referred to Ref. [16].

In the fabrication of inversion mode MOSFETs, oxidation was performed on p-type ((1.5–3.0)E17 Al/cm^3) β-SiC (on Si) samples at 1473K for 2.16E3s to produce 50 nm of gate oxide. Poly-Si (530 nm) was then deposited via low pressure CVD at 893K and degenerately doped via P diffusion at 1173K for 300s from a glass layer subsequently removed. A circular gate pattern was applied via photo resist and the poly-Si stripped from all remaining area. A dual N implant (5E14/cm^2 at 70 keV and 3.35E14/cm^2 at 40 keV) at 873K produced a dual peak concentration of 5E19/cm^3; the poly-Si acted as an implant mask thus producing a self-aligned gate. The source and drain areas were patterned in dark field and etched with buffered HF for 120 s. Following deposition of 200 nm of $TaSi_2$, the remaining photo resist was stripped and the samples cleaned, heated in (1) vacuum for 300 s to anneal the $TaSi_2$ contacts and (2) forming gas (8%H_2/92%N_2) at 723K for 300 s to anneal the surface states at the oxide/SiC interface.

Depletion mode MOSFETs were fabricated in 1.2 μm thick n-type layers on ≈0.5 μm thick p-type β-SiC thin films grown by CVD at 1773K on the Si (0001) face of 6H α-SiC crystals. The source of p-type doping (p=1.6E/16cm^3) was Al which diffused from the heavily Al doped 6H α-SiC substrate. These acceptor species also highly compensated the n-type layers, thus producing the unusually low carrier concentration. The resulting films were polished, oxidized, etched, cleaned, received a deposition of P-doped poly-Si and patterned to form the gate contacts similarly to the preparation of the inversion mode devices. Two different gate lengths of 7.2 μm and 2.4 μm were employed. The n^+ source and drain areas were subsequently formed by dual N^+, implants using the same dose and energy parameters employed in the inversion devices. The temperature of the implants was 773K. Windows for source and drain contacts, the $TaSi_2$ contacts and their anneal and the annealing of the total device were also conducted similarly to that described for the inversion devices.

4. RESULTS and DISCUSSION

The β-SiC layer produced by chemical conversion of the Si (100) surface was monocrystalline and microscopically rough, varying in thickness from 5–12 nm. It contained a high density of defects that were primarily {111} microtwins and intrinsic stacking faults. In addition, localized regions of the converted layer exhibited disorder. The microtwins and stacking faults were also identified in the subsequently deposited β-SiC films. Their density was higher at the interface but decreased over a distance of 3–4 μm from the interface; it became approximately constant to the surface, even for 20 μm thick films. An additional defect, namely APBs, also occurred in the films grown directly on Si (100). We have found that the APBs can be completely eliminated if the films are deposited on 4° off-axis Si (100) with the [100] inclined toward [011]. For additional information regarding these results, the reader is referred to Refs. [15–17] and to the paper by Pirouz [this volume].

The films grown directly on the C (000$\bar{1}$) and the Si (0001) faces of the α-SiC substrates were always β-SiC (111). The surface of the film grown on (0001) was very smooth and reflective whereas that grown on (000$\bar{1}$) was relatively rough and unsuitable for device fabrication. In contrast to films grown on the Si substrates, no defects were observed in these films when examined by cross sectional transmission electron microscopy (XTEM). High resolution XTEM shows an abrupt and coherent β-SiC/α-SiC interface (Fig. 1). A single atomic layer runs completely across the interface which indicates that the growth direction of the β-SiC films on the α-SiC C face was exactly [$\bar{1}\bar{1}\bar{1}$]. However, examination in plan view revealed the presence of double positioning (DP) boundaries. In

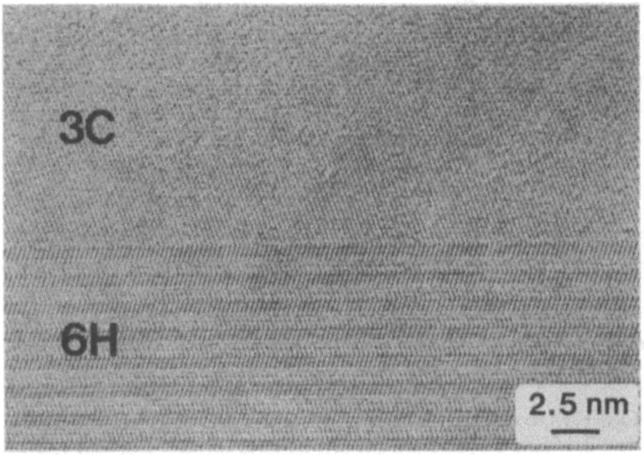

Figure 1. High-resolution XTEM micrograph of the β– SiC/6H–SiC interface (Coutesy of S. Nutt, Brown University, Providence, RI).

these defects, the material enclosed by the boundaries is oriented 60° to the surrounding matrix around <111>. Recent research has employed α-SiC crystals with the [000$\bar{1}$] inclined 3° toward [11$\bar{2}$0] heated to 1773K for 600 s. The DP boundaries were eliminated by this procedure, and a mirror smooth film was produced. However, the resulting film was the 6H α-SiC polytype. Plan view TEM showed high structural quality of the 6H α-SiC epilayer (Fig. 2); the irregular bends in this micrograph are strain contours which are believed to be caused during ion milling of the TEM sample. As a consequence of the very recent nature of this study utilizing α-SiC substrates, the results of most of the remaining research reported below are concerned with β-SiC grown on Si.

Ion implantation at room temperature and at liquid nitrogen (LN) conditions was investigated previously [10]. However, little success was achieved in affecting precipitate- and/or damage-free solid phase epitaxy (SPE) regrowth in the β-SiC films previously amorphized by ion implantation. In addition, no pronounced changes in the electrical properties of any of these films occurred until anneal temperatures equal to or in excess of 1673K were employed. As such, high temperature implantation is currently being studied to surmount some of these problems. To date, we have learned that implantation of Al or N at 623K, 823K, and especially 1023K allows sufficient *in situ* annealing during implantation to produce both essentially damage-free implant regions as well as virtually complete electrical activation of the implanted species [16]. The backscattering spectra from the 1023K implants nearly coincide with the aligned spectrum of the unimplanted material. In addition cross-sectional TEM analysis revealed that implantation at this highest temperature resulted in neither visible lattice damage nor precipitates. Finally, it should be noted again that a slight increase in the electrical activation of the high temperature implants can be achieved if the samples are annealed at 1473K for 1800 s.

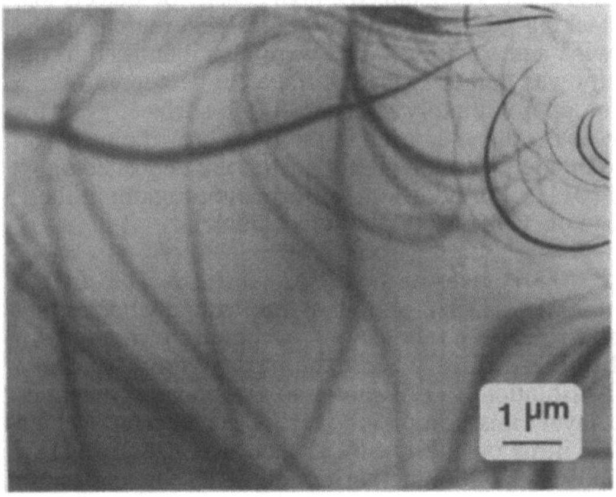

Figure 2. Plan view TEM micrograph of a 6H-SiC thin film grown on 3° off 6H-SiC (000$\bar{1}$) C face towards [11$\bar{2}$0]. Electron beam was parallel to [0001]. Note no DPBs in the film.

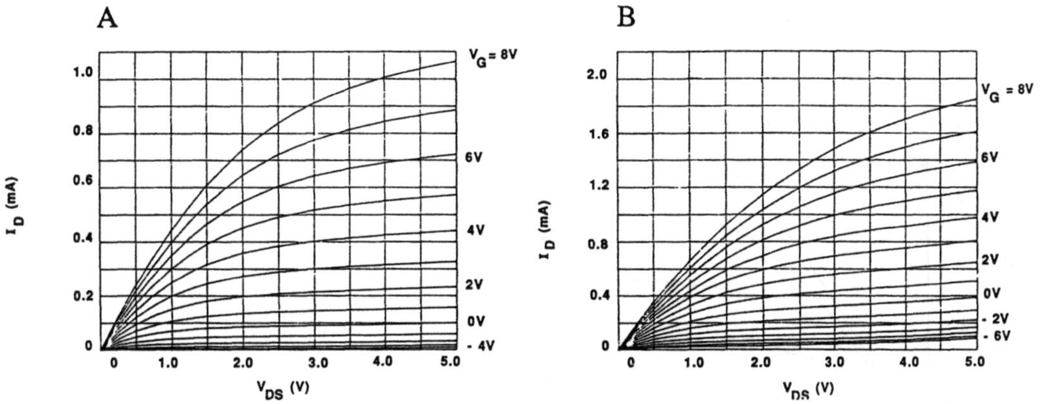

Figure 3. Current-voltage characteristics of a β-SiC n-channel MOSFET with a channel length and channel width of 20 μm and 421 μm, respectively, at (A) 300K and (B) 673K.

Figure 3(A) shows the room temperature I_D vs V_D characteristics for an inversion-mode MOSFET having a 20 μm gate length over the gate voltage range of -4.0 to +8.0 V. The leakage currents at V_{DS} = 5 V and V_g = -7 were <10 μA and the threshold voltage and maximum transconductance were -2.4 volts and 0.335 mS/mm, respectively. This device also showed good stability as a function of temperature to the highest measured point of 673K, as shown in Fig. 3(B). However, the leakage current increased steadily with temperature to a maximum of 85 μA at 5 V and 673K. The transconductance increased to 0.421 mS/mm and the threshold voltage decreased to –5 V. This negative shift in threshold voltage with temperature was caused by a change in the barrier height between the metal and the semiconductor.

The drain current-voltage characteristics of a depletion-mode MOSFET in β-SiC (111) at 296K are shown in Fig. 4(A). This particular device had a gate length and width of 7.2 μm and 390 μm, respectively. The source contact to drain contact distance was 24 μm. The device showed very stable drain current saturation out to a drain-source voltage of 25 V. (This trend actually continued to V_{DS}≈30 V, at which point the oxide underwent breakdown.) The threshold voltage was determined at a gate voltage, V_G, of -12.9 V from a plot of $\sqrt{I_{DS(sat)}}$ versus V_G. The leakage current at V_{DS}=25 V in the "off" state (V_G=-15 V) in this device was 3.75 μA. The negative threshold voltage was due, in part, to the voltage required to deplete the n-channel, but it was made more negative by the presence of either fixed oxide charge or mobile ion charge in the range of 5–6x10^{12}/cm^2, as obtained from MOS C–V curves of the gate oxide. The 723K anneal in forming gas did not change the threshold voltage. The maximum transconductance of this device at room temperature with V_{DS} fixed at 20 V was 5.32 mS/mm at V_G=2.5 V.

When this device was heated to 573K and allowed to stabilize for 15 mins., the transconductance actually increased to a maximum of 6.00 mS/mm at V_G=5.5 V and V_{DS}=20 V. This trend can also be observed by the increased current at V_G=0 V in the 573K drain characteristics shown in Fig. 4(B). The reason for this increase is currently under investigation. Despite the increase in temperature, the drain current saturation was still very stable to V_{DS}=25 V. The leakage current at V_{DS}=25 V and V_G=15 V increased to 22 μA, and the threshold voltage shifted negatively to V_G=13.3 V.

The drain characteristics of this depletion device were also determined at 673K and at every 50K interval to a maximum of 923K. Results for this last temperature are shown in Fig. 4(C). The transconductance remained unchanged at 673K; however, it decreased with further increases in temperature. The maximum transconductance measured was 11.9 mS/mm in a 2.4 μm gate length device at 673K. The lower transconductance of the device in Fig. 4(C) at 923K is demonstrated by the lower current at V_G=0 V, as compared with the previous curves. Although the transconductance at this temperature became very erratic above V_G=1 V, it reached a maximum of about 4.8 mS/mm at V_G=8 V and V_{DS}=20 V. The decrease in transconductance above 673K is due to increasing lattice scattering with temperature. The threshold voltage again shifted negatively to V_G=14.8 V at 923K. The leakage current increased to 128 μA at V_G=-15 V and V_{DS}=25 V. When the temperature was raised to 973K, this device showed similar current saturation, but the gate oxide experienced breakdown. Thus, current was being injected at the gate and the device could not be cut off.

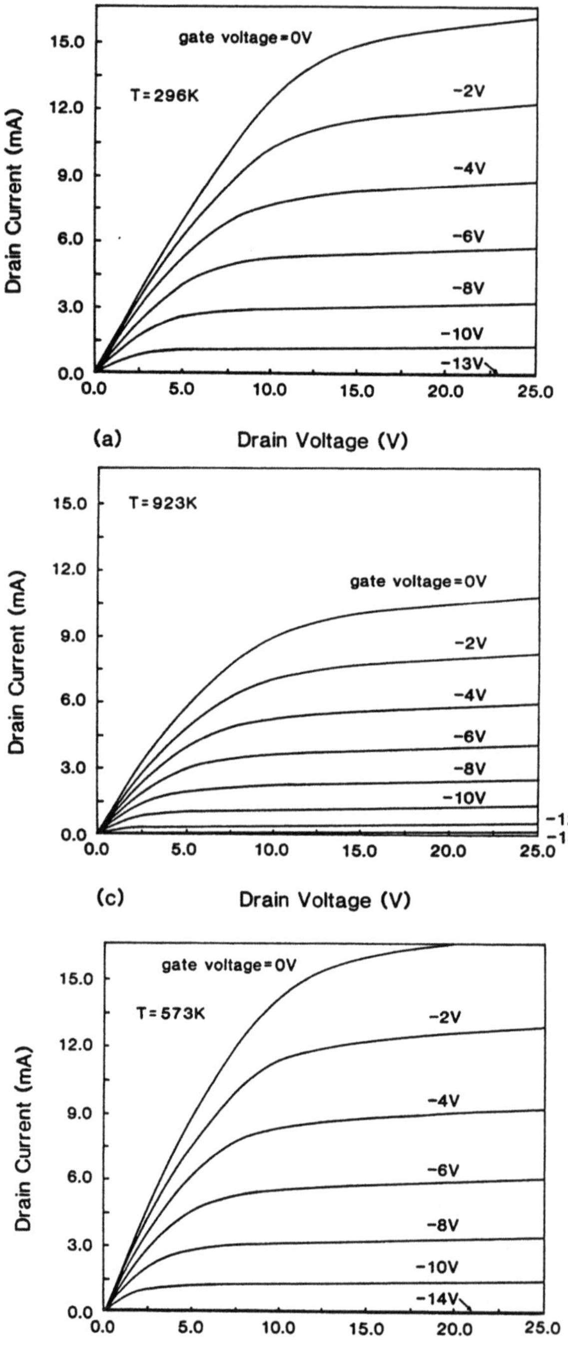

Figure 4. Drain characteristics of depletion mode n-channel MOSFET in β-SiC (111)thin film at (A) 296K, (B) 573K, (C) 923K. The gate length and width of this device were 7.2 μm and 390 μm, respectively.

5. CONCLUSIONS

Monocrystalline thin films of high purity β-SiC have been grown directly on Si (100) and α-SiC (0001). Problems derived primarily from interface related defects and subboundaries in the material have prevented the optimization of the quality of the films. However, deposition on these materials oriented 2–4° off-axis allows the elimination of the APBs in the material grown on Si (100) and the DPBs in the material deposited on SiC (0001); the latter resulting combination of α-SiC on α-SiC appears especially viable for device fabrication. Despite the defects, the existence of the films has allowed substantial progress in the development and understanding of doping procedures, dopant activation, electrical properties and device development. This collective research has shown that continued emphasis must be placed on improvement of the quality of the films.

6. ACKNOWLEDGEMENTS

The authors enthusiastically acknowledge the support of this program by the Office of Naval Research under contract N00014-82-K-0182P005. Appreciation is also expressed to Dr. S. P. Withrow for his assistance with the ion implantation research under U.S. D. O. E. Division of Materials Sciences contract D.E.-AC05–840–R21400 with Martin Marietta Energy Systems, Inc., and to Dr. G. Carver of NBS and J. Wortman of NCSU for valuable discussions regarding device research.

7. REFERENCES

1. On assignment from Kobe Development Corporation, Research Triangle Park, NC 27709.
2. H. J. Kim, R. F. Davis, X. B. Cox and R. W. Linton, Physical and Chemical Nature of Films Formed on Si (100) Surface Subjected to C_2H_4 at Elevated Temperatures, J. Electrochem. Soc., **134,** 2269 (1987).
3. S. Nishino, J. A. Powell and H. A. Will, Appl. Phys. Lett. **42**, 460 (1983).
4. A. Suzuki, K. Furukawa, Y. Higashigaki, S. Harada, S. Nakajima and T. Inoguchi, J. Cryst. Growth **70**, 287 (1984).
5. A. Addamiano and P. H. Klein, J. Cryst. Growth **70**, 291 (1984).
6. K. Sasaki, E. Sakuma, S. Misana, S. Yoshida and S. Gonda, Appl. Phys. Lett. **45**, 72 (1984).
7. P. H. Liaw and R. F. Davis, J. Electrochem. Soc. **132**, 642 (1985).
8. P. Rai-Choudhury and N. P. Formigoni, J. Electrochem Soc. **116,** 1440 (1969).
9. I. Berman, C. E. Ryan, R. C. Marshall and A. Littler, AFCRL report #72-0737, Dec. 19, 1972.
10. J. A. Edmond, H. J. Kim and R. F. Davis, in Rapid Thermal Processing, ed. by T. O. Sedgwick, Thomas E. Seidel and B.-Y. Tsaur, Vol. 52 (MRS, Pittsburgh, 1986), p.157.
11. K. Shibahara, T. Saito, S. Nishino and H. Matsunami, in Extended Abstracts of the 18th (1986 International) Conference on Solid State Devices and Materials, 1986, p.717.
12. Y. Kondo, T. Takahashi, K. Ishi, Y. Hayashi, E. Sakuma, S. Misawa, D. Daiman, M. Yamanaka and S. Yoshida, IEEE Electron Device Letters, **EDL-7,** 404 (1986).
13. K. Shibahara, T. Saito, S. Nishino and H. Matsunami, in Extended Abstracts of the 18th (1986) International Conference on Solid State Devices and Materials, 1986, p. 717.
14. H. S. Kong, Y. C. Wang, J. T. Glass and R. F. Davis, accepted for publication by J. Mat. Res., 1988.
15. H. S. Kong, J. T. Glass and R. F. Davis, Appl. Phys. Lett. **49**, 1074 (1986).
16. J. A. Edmond, S. P. Withrow, W. Wadlin and R. F. Davis, in Interfaces, Superlattices and Thin Films, ed. by J. D. Dow, I. K. Schuller and J. Hilliard, Vol. 77 (MRS, Pittsburgh, 1987) p.193.

Plasma-Assisted Chemical Vapor Deposition of Cubic Silicon Carbide on Silicon Substrate

S. Nishino and J. Saraie

Department of Electronics and Information Science,
Faculty of Engineering and Design, Kyoto Institute of Technology,
Matsugasaki, Sakyo-ku, Kyoto 606, Japan

Cubic SiC films were deposited on Si substrates by plasma-assisted chemical vapor deposition (CVD) using Si_2H_6 and C_2H_2 as source materials at a substrate temperature of 800 - 1150 °C. The activation energy for SiC formation with the plasma was lower (11.4 kcal/mole) than that without the plasma (33.7 kcal/mole). The film was polycrystalline at a substrate temperature less than 1000 °C, however, highly oriented crystalline films were obtained at 1150 °C. I-V curves of heterojunction diodes (n-SiC/p-Si) showed good rectifying characteristics with a value of 1.5-2 for the ideal factor n. The temperature dependence of the I-V curve was also studied.

1. Introduction

Silicon carbide is a promising material for electronic and optical devices under severe conditions, such as high temperature and strong irradiation environments, because of its large energy gap, temperature stability, and chemical inertness. Single crystal growth of cubic SiC on Si was difficult for a long time because of large lattice mismatch (20%) between SiC and Si. Recently, techniques for epitaxial growth of SiC have been estblished and many device characteristics have been reported.[1,2,3,4] The temperature of crystal growth, however, is high, being around 1300 °C. In high temperature growth, slip lines appear on the Si substrates and dislocations are induced by large thermal stress[5]. From the viewpoint of device fabrication, low temperature growth is desired. Plasma assisted CVD is widely used for the deposition of amorphous Si[6] ,Si-C alloys[7] and micro-crystalline Si[8] to control the composition and bandgap of the film. For single-crystal growth, several promising results are reported using plasma-assisted CVD for Si[9], III-V[10] and II-VI[11] compound semiconductors, however, it has not been well studied for SiC. We examined crystal growth of cubic SiC at a low temperature employing RF plasma excitation and gas sources of Si_2H_6 and C_2H_2. Heterojunction properties of n-SiC/p-Si were also studied.

Springer Proceedings in Physics, Vol. 34 Amorphous and Crystalline Silicon Carbide
Editors: G.L. Harris and C.Y.-W. Yang

2. Experimental

Cubic SiC films were deposited on silicon(100) substrates using plasma-assisted CVD and Si_2H_6 and C_2H_2 as source materials. A schematic diagram of the CVD system is shown in Fig.1. A horizontal quartz tube, 30 mm ID and 500 mm long, was used for the reactor. The susceptor was made of SiC coated graphite (20x40x7 mm^3), and heated by a 2kW IR lamp with an elliptical reflector. The substrate temperature was measured by a thermocouple inserted in the susceptor. The plasma was excited by RF of 13.56 MHz using two electrodes 20 cm apart and attached around the reaction tube as shown in Fig.1. The flow rates of the gases were controlled by mass flow controllers (Nippon Tylan Corp.). Pressure was measured using a Pirani gauge.

Crystal growth of cubic SiC on Si substrates consisted of three steps: etching, carbonization, and CVD growth, as shown in Fig.2. The entire system was evacuated by a rotary pump and backfilled with Ar. After 10 min of Ar flow, pressure of the reaction tube was adjusted to 4 Torr. The substrate was heated externally by IR lamp irradiation and kept at 1000 °C for 5 min for plasma etching at a power of 100 W for 5 min. The carrier gas was subsequently changed from Ar to H_2 and the pressure readjusted to 4 Torr. The flow rate of $C_2H_2 \sim 0.1$ sccm was then employed for 4 min and the carbonization procedure conducted

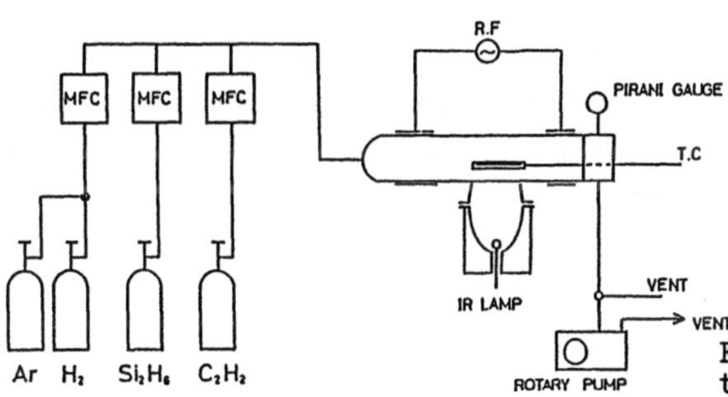

Fig.1 A schematic diagram of the CVD system.

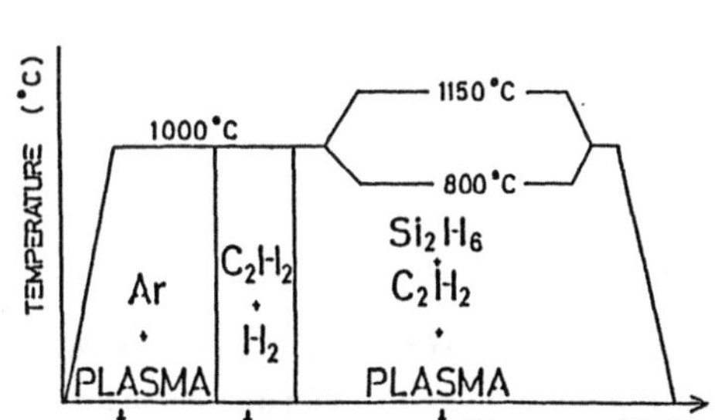

Fig.2. Temperature program of CVD: t_e, t_c, and t_{cvd} mean the time for etching, carbonization, and CVD growth, respectively.

without the plasma. Finally, the reaction gases of Si_2H_6 and C_2H_2 were added to the plasma excited hydrogen carrier and crystal growth commenced. Crystal growth was carried out at a substrate temperature in the range of 800-1150 °C with the standard gas composition: H_2~200 sccm, Si_2H_6~0.1 sccm, C_2H_2~ 0.1 sccm. Other growth conditions included a Si/C(gas flow ratio)=1, a pressure of 4 Torr, and RF power of 50 W. The deposition time was usually 20-60 min. The thickness of the deposited cubic SiC film was determined by optical reflection mesurements.

3. Result and Discussion

The layers grown on the Si substrate in the temperature range of 800-1150 °C with the standard gas composition showed different colors which depended on their thickness. In the above temperature range the growth rate is expressed by an exponential function of the reciprocal substrate temperature, which indicates that the growth is limited by surface reaction, as shown in Fig.3. In the figure, squares show the deposition with plasma and circles that without plasma. The activation energies for the formation of cubic SiC with and without plasma are 11.4 kcal/mole and 33.7 kcal/mole, respectively. In plasma assisted CVD, the hydrogen radicals might reduce the barrier height which is associated with the surface migration of active chemical species by covering the growth surface.[12)]

The carbonized Si surface, a buffer layer, was examined by reflection electron diffraction (RED). The surface was polycrystalline even though parameters such as the flow rate of C_2H_2, substrate temperature, and reaction time, were varied.

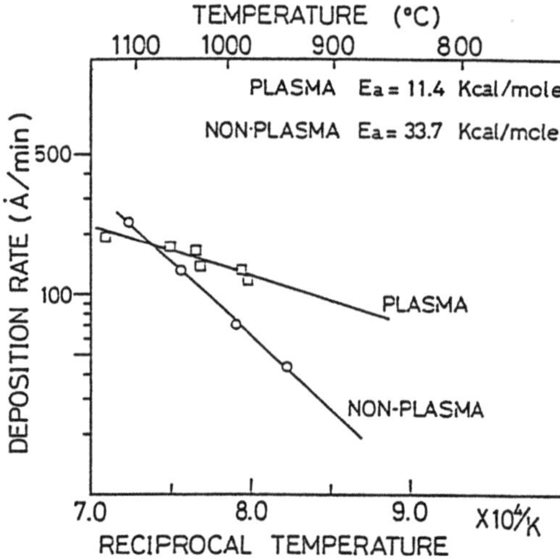

Fig.3. Temperature dependence of the growth rate. Squares and circles show deposition with and without plasma, respectively.

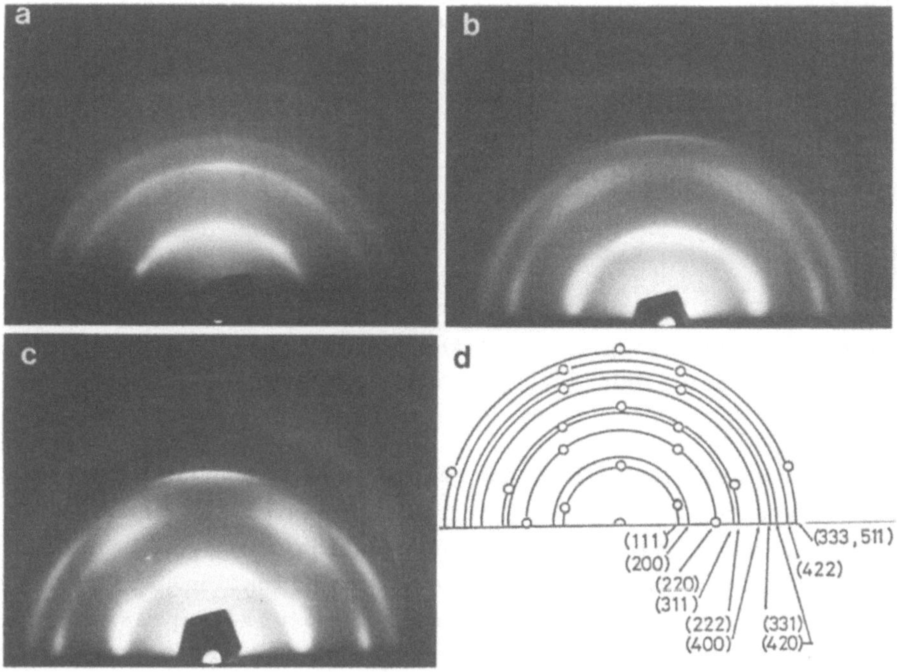

Fig.4. Reflection electron diffraction patterns for the film prepared at different substrate temperatures:
(a) 1000 °C,(b)1100 °C,(c)1150 °C,(d)indexed pattern of (c).

The growth temperature is one of the most important parameters, in the determination and control of the crystal quality, surface morphology and growth rate. The RED patterns of the grown layers are shown in Fig.4. The diffraction pattern of the film prepared at 1000 °C and presented in Fig.4(a) shows clear Debye rings which indicate a polycrystalline structure. Complete Debye rings split into discrete rings, which indicate an oriented structure in the film, when the substrate temperature became higher than 1050 °C, as shown in Fig.4(b). At a substrate temperature of 1150 °C, a highly oriented film was obtained, as shown in Fig.4(c). A schematic representation of the diffraction pattern when the [111] axis of the crystallites in the SiC film is normal to the substrate surface is shown in Fig.4(d).

The substrate temperature dependence of the surface morphology is shown in Fig.5. At the lower substrate temperature, the surface was smoooth, however, the surface became rougher at the higher substrate temperature because of a low nucleation rate.

Heterojunction properties of n-SiC/p-Si were also studied. The Si substrates were p-type with a resistivity of 5-8 ohm.cm. The cubic SiC

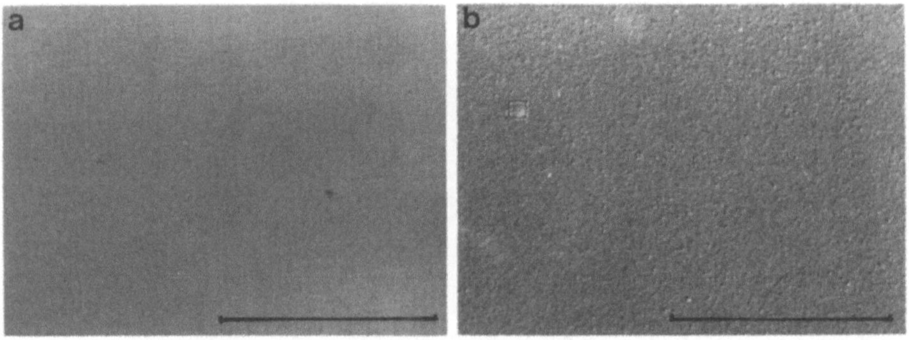

Fig.5. Nomarski microphotographs of the grown layer for different substrate temperatures:(a)900 °C,(b)1150 °C. Scale is 50 µm.

films (thickness about 3000 A) on Si showed n-type conduction with a resistivity of about 0.2 ohm.cm and carrier concentration of $10^{18}cm^{-3}$. Ohmic contacts were made by vacuum evaporation using AuGa for p-Si and Al for n-SiC. The diodes showed good rectification. Current(I)-voltage(V) curves measured at room temperature were expressed by a relation of $I \propto \exp(eV/nkT)$, where n-values (ideal factor) were 1.5 to 2 when the diodes were prepared at 1100°C. The I-V characteristics of the diode were recorded at various temperatures, as shown in Fig.6. In this figure the dotted lines show the current by subtracting a component of series resistance of the diode. The slopes of these semilog plots have two regions. One is expressed as $I \propto \exp(BV)$, indicating tunnel current, in which B is a constant independent of temperature, and the other is $I \propto \exp(eV/nkT)$. At lower voltage, the value B was almost constant with values in the range of 23-24 , indicating a tunneling current is dominant. At high voltage, the n-value changed from 2.1 to 1.5 with increasing the temperature, indicating that the current is dominated by generation-recombination.

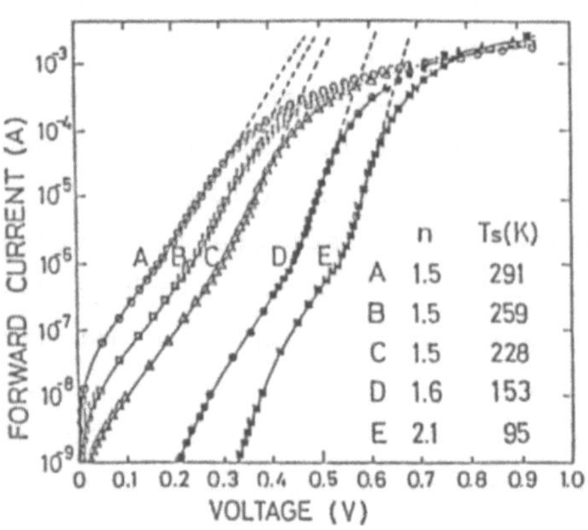

Fig.6. Temperature dependence of I-V characteristics for n-SiC/p-Si heterojunction.

Recombination and tunneling currents may be due to the interface states produced by large lattice mismatch between SiC and Si.

Capacitance(C) versus voltage(V) characteristics of the diodes were measured, and the relation $V \propto 1/C^2$ was obtained, which indicates that the junction was abrupt. The diffusion voltage obtained in the C-V characteristics was about 0.6 V, this value is slightly larger than that obtained from the I-V characteristics.

4. Conclusions

Plasma excitation is effective in decreasing the activation energy for growth of cubic SiC. Highly oriented cubic SiC is deposited at a substrate temperature of 1150 ^{o}C. Heterojunctions of n-SiC/p-Si had abrupt junctions. The temperature dependence of I-V characteristics of these junction was studied.

The authors express their appreciation to Mr.T.Fukada for the film deposition and to Mr.N.Kitamura for the electrical measurements of the diode.

References

1) S.Nishino, J.A.Powell and H.Will;Appl.Phys.Lett.**42**,460(1983).

2) S.Nishino, H.Suhara, H.Ono and H.Matsunami; J.Appl.Phys.**61**,4889(1987).

3) K.Shibahara, S.Nishino and H.Matsunami; Appl.Phys.Lett.**50**,1888(1987).

4) K.Shibahara, T.Saito, S.Nishino and H.Matsunami; IEEE Electron Devices **EDL-7**,692(1986).

5) H.P.Liaw and R.F.Davis; J.Electrochem.Soc.**131**,3014(1984).

6) W.E.Spear and P.G.Lecomber; Philos.Mag.**33**,935(1976).

7) J.Saraie, Y.Fujii, M.Yoshimoto, K.Yamazoe and H.Matsunami; Thin Solid Films **117**,59(1984).

8) A.Matsuda, T.Yoshida, S.Yamasaki and K.Tanaka; Jpn.J.Appl.Phys.**20**,L439(1981).

9) K.Nagamine, A.Yamada, M.Konagai, and K.Takahasi, Jpn.J.Appl.Phys.**26**,L951(1987).

10) A.D.Huelsman,R.ReifandC.G.Fonstad; Appl.Phys.Lett.**50**,206(1987).

11) T.Hariu, S.Yamaguchi, S.Goto, K.Shida and K.Matsushita; Ext. Abst. 18th Conf.Solid State Devices and Materials,**Tokyo**,639(1986).

12) K.Matsuda; J.Non-Cryst.Solid **59&60**,767(1983).

Reactive Ion Etching for SiC Device Fabrication

W.-S. Pan and A.J. Steckl

Center for Integrated Electronics and Department of Electrical, Computer and Systems Engineering, Rensselaer Polytechnic Institute, Troy, NY 12181, USA

Reactive ion etching of SiC thin films in a variety of fluorinated gases (SF_6, $CBrF_3$ and CHF_3) has been investigated. The plasma conditions are chosen to achieve the selective SiC to Si and SiO_2 etching and anisotropic patterning of SiC thin films needed for device fabrication. By using $CBrF_3$/75%O_2 and CHF_3/90%O_2 plasmas at 200W and 20mTorr, we have obtained a SiC:Si etch rate ratio (i.e. selectivity) of 2:1 in both gases. At a corresponding SiC etch rate of approximately 40nm/min under these conditions, tapered edge profiles were found with a vertical-to-lateral anisotropic ratio of 12.5:1 and 6.3:1, respectively. The maximum SiC etch rate of 53.3nm/min was obtained in SF_6/35%O_2 and resulted in an anisotropic ratio of over 10:1. The highest SiC:Si reverse selectivity of 1:44 was measured for SF_6/10%O_2 at 200W and 20mTorr. SiC to SiO_2 etch rate selectivity has also been obtained. For example, using CHF_3/80%O_2 plasma a SiC:SiO_2 selectivity of 1.3:1 was measured. The optimum SiC:SiO_2 reverse selectivity of 1:3.5 was obtained in pure CHF_3 plasma. The roles of fluorine and oxygen in determining the etch rate selectivity are discussed.

1. Introduction

We have previously reported [1,2] on the first evidence of plasma-assisted etching of SiC thin films using fluorinated gases. To fully utilize plasma-assisted etching for SiC technology, one needs to explore and understand the effect of various plasma parameters including pressure, power, etching species, dilutants, etc. The use of CF_4[3,4], CF_4/O_2[1,5,6], NF_3[4] and SF_6[7] plasmas to pattern SiC has been previously reported. However, in order to incorporate SiC films into Si VLSI technology, a selective, and frequently anisotropic, etching process with respect to Si and SiO_2 is required[2]. For example, device fabrication considerations require SiC etching selectivity and anisotropy to define the active area well in the SiC/Si heterojunction device, as shown in Fig.1a. For the SiC homo-device (MOSFET) shown in Fig. 1b, a selectivity of SiO_2 to SiC higher than unity is required for gate

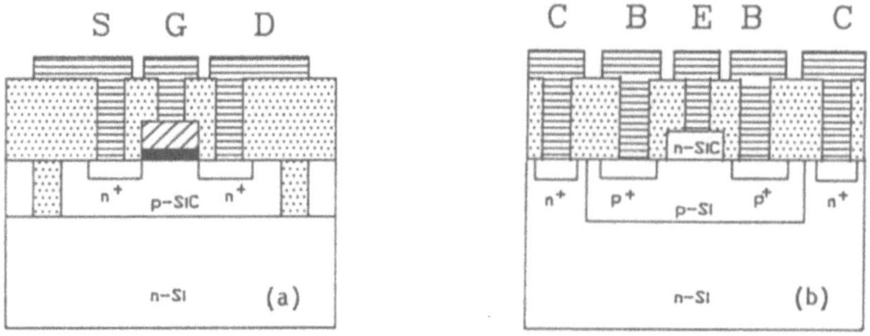

Fig. 1.(a) Device structure for SiC/Si heterojunction transistor. (b) Homo-device structure of SiC MOSFET

Springer Proceedings in Physics, Vol. 34 **Amorphous and Crystalline Silicon Carbide**
Editors: G.L. Harris and C.Y.-W. Yang

area patterning and an anisotropic isolation etching process is needed for high density device applications. Therefore, in this paper , we have explored various fluorinated gas mixtures, such as $CBrF_3/O_2$, SF_6/O_2 and CHF_3/O_2, which can begin to fulfill these requirements.

2. Experiments

SiC thin films were deposited by RF sputtering onto silicon substrates in a planar system. The film preparation and apparatus used in these experiments have been previously reported [1,2]. The films were furnace annealed at 1100°C in a nitrogen ambient for 30 minutes. N-type (4-6 ohm-cm) (100) Si wafers were used to determine the Si etch rate. Oxidized silicon substrates (in steam at 1100°C) were used for SiO_2 etching. The etching experiments were carried out in a parallel plate reactor [1,2] equipped with a computer-controlled grating monochromator for measuring optical emission within the plasma. The optical emission intensity of various species in the plasma is roughly proportional to their concentrations and has been used in this work to investigate the relationship between emission intensity and process. For a more accurate relationship between emission intensity and plasma concentration an actinometry calibration [8] can be performed. The DC self-bias of the RF electrode was also monitored. To determine the etch rate in various ambients, Al was used as a thin film mask since it is suitable for both low and high percentage of oxygen. The etching profile has been measured by profilometer and SEM.

3. Results

The etch rates were determined as a function of oxygen percentage in SF_6, $CBrF_3$ and CHF_3 gas plasmas. In Fig. 2a, the SF_6/O_2 plasma is investigated and the SiC, Si and SiO_2 etch rates are shown as a function of oxygen percentage (from 0% to 90%) at a pressure of 20mTorr, a total flow rate of 20sccm and power of 200W (or $0.42W/cm^2$). In Fig. 2b are shown the corresponding DC self-bias and the relative intensity of fluorine [F] and oxygen [O] emission at wavelengths of 703.7 and 780 nm, respectively. The highest SiC etch rate measured is 53.3nm/min at 35% O_2, where the etch rates of Si and SiO_2 are 1.4μm/min and 60nm/min. The highest reverse selectivity of SiC to Si obtained is 1:44 for SF_6/10%O_2, and the corresponding etch rates are 45nm/min and 2μm/min. In Fig. 3a, the results obtained using $CBrF_3$ and O_2 mixtures in a similar etching experiment are shown. A SiC to Si etch rate ratio of 2:1 was reached at 75%O_2, where the etch rates of SiC and Si are 37.5nm/min

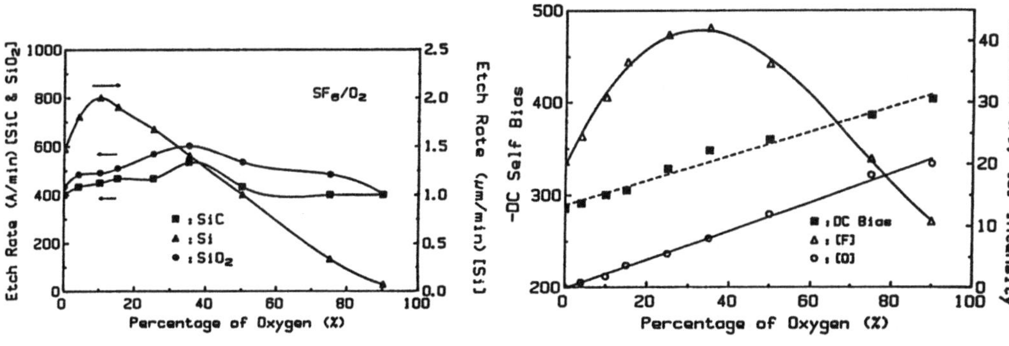

Fig. 2.(a) Etch rates of SiC, Si and SiO_2, (b) DC self bias and fluorine, oxygen emission intensity as a function of O_2 in SF_6 + % O_2 at 20sccm, 200W, 20mTorr

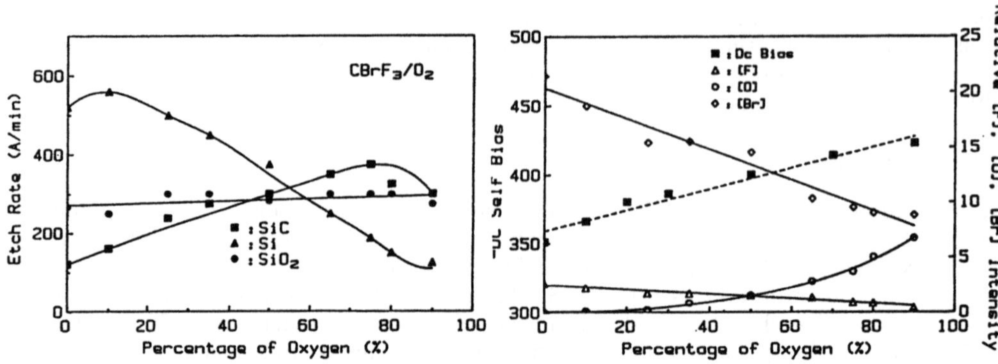

Fig. 3.(a) Etch rates of SiC, Si and SiO_2 , (b) DC self bias and fluorine, oxygen, bromine emission intensity as a function of O_2 in $CBrF_3$ + % O_2 at 20sccm, 200W, 20mTorr

and 18.8nm/min, respectively. The DC bias and [F], [O] and [Br] emission intensity (the later at a wavelength of 336nm) are presented in Fig. 3b. The etching data obtained for CHF_3 and O_2 mixture are shown, under the same plasma conditions as previous experiments, in Fig. 4a. The corresponding DC bias, [F], [O] and [H](486nm) emission intensity are shown in Fig. 4b. A maximum etch rate ratio of SiC to Si of 2:1 was obtained for CHF_3/90%O_2. In a pure O_2 plasma, the etching rate is almost zero for all samples. The edge profiles obtained using these fluorinated gas were examined by SEM. In general, no undercut was observed, but rather a tapered profile of the SiC film edge was found . In the microphotograph of Fig. 5, it can been seen a vertical-to-lateral profile ratio of 10.3:1 has been obtained in an SF_6/35%O_2 plasma. The Si substrate underneath the SiC film was removed by the isotropic etching of Si. A tapered profile ratio of 6.3:1 was obtained by using $CBrF_3$/75%O_2 plasma as shown in Fig.6. In Fig.7, it is shown that with a CHF_3/90%O_2 mixture the highest anisotropic SiC etching ratio of 12.5:1 is obtained and that no undercut of either the SiC or Si is evident. In this microphotograph, the SiC film is shown with Al thin film mask still present.

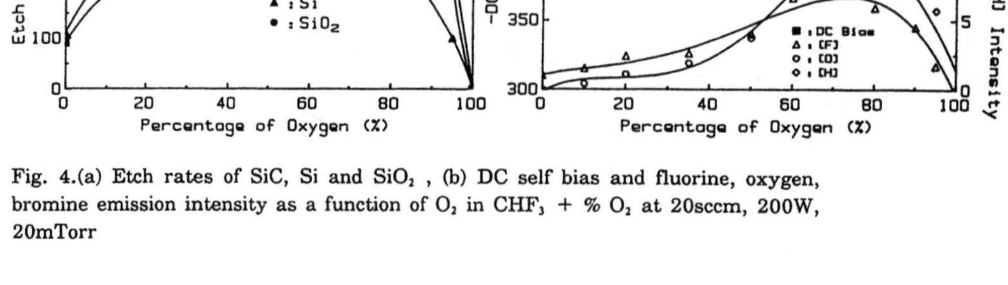

Fig. 4.(a) Etch rates of SiC, Si and SiO_2 , (b) DC self bias and fluorine, oxygen, bromine emission intensity as a function of O_2 in CHF_3 + % O_2 at 20sccm, 200W, 20mTorr

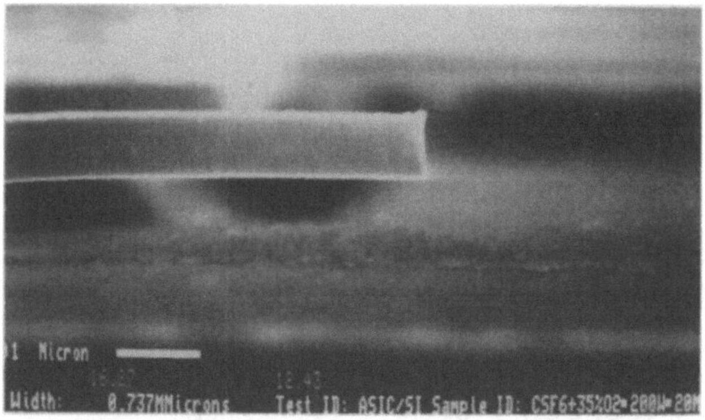

Fig. 5. SEM photomicrograph of SiC layer on Si(100), RIE-etched by $SF_6/35\%O_2$, 20sccm,20mTorr, 200W. (Si was overetched and SiC layer was floated)

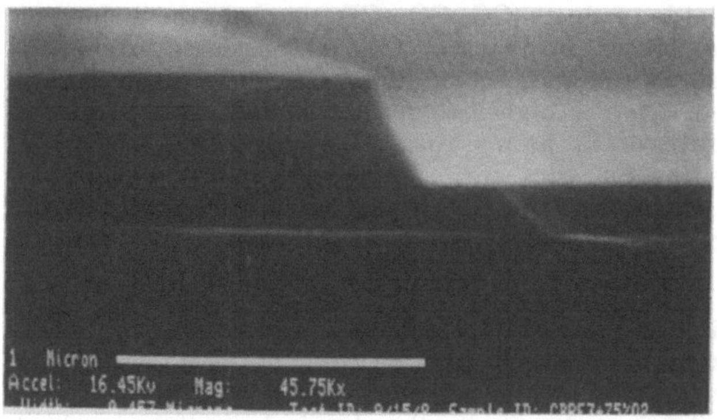

Fig. 6. SEM photomicrograph of SiC layer on Si(100), RIE-etched by $CBrF_3/75\%O_2$, 20sccm,20mTorr, 200W (etching was stopped inside SiC layer)

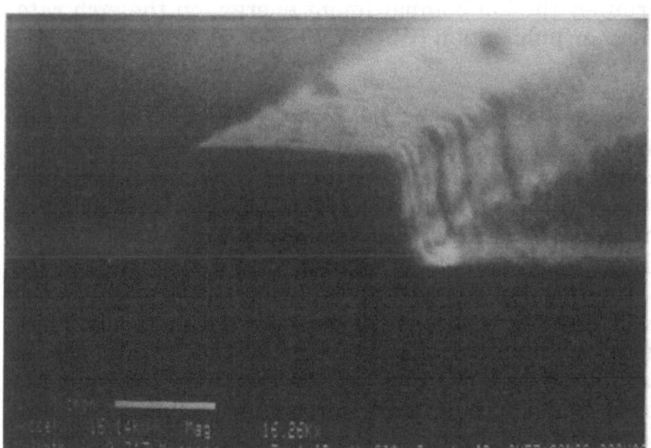

Fig. 7. SEM photomicrograph of SiC layer on Si(100), RIE-etched by $CHF_3/90\%O_2$, 20sccm,20mTorr, 200W (sample was etched into Si-substrate, Al-mask still in place

4. Discussion

Reactive ion etching consists of two basic processes : chemical and physical mechanisms. For SiC and Si etching in fluorinated gases, the most likely chemical reactions are given below :

$$\mathbf{Si + 4F \dashrightarrow SiF_4} \quad (1)$$

$$\mathbf{C + xO \dashrightarrow (CO, CO_2)} \quad (2)$$

$$\mathbf{C + xF \dashrightarrow CF_x} \quad (3)$$

$$\mathbf{SiC + xF + yO \dashrightarrow SiF_4 + (CO, CO_2) + CF_x} \quad (4)$$

In reaction (1), Si is removed mainly by reaction with fluorine to form SiF_4 volatile gas. Carbon can react with either oxygen or fluorine atoms, as indicated in reactions (2) and (3), and form different volatile products (CO, CO_2, CF_X, etc.). In the etching of SiC a two-step etching process, shown in reaction (4), is considered to be an appropriate etching model, wherein silicon and carbon atoms are removed separately. Under most conditions, the Si etch rate is higher than that of SiC. This is due to comparatively higher removal rate of Si atoms than C atoms. As a result, a carbon-rich surface is formed on the SiC film during etching, becoming a potential rate-limiting step under certain conditions. The C-rich surface has been verified by Auger electron spectroscopy in previous work [1,2] as well as this work.

In general, selective etching is the result of chemical, rather than physical, processes. Therefore in this work we have concentrated on studying those parameters (i.e. the etching species concentrations in the plasma) which will allow us to understand the effect of various fluorinated gases on the chemical reactions involved in etching SiC. At the same time, it is important to bear in mind that the physical processes also play an important role in determining the SiC etch rate. This has been clearly shown previously for CF_4[1], SF_6[1] and CHF_3[2] reactive ion etching. The effect of the DC bias, which determines the ion bombardment energy, on the etch rate will, however, be the subject of another publication [9].

To obtain selective SiC to Si etching, one needs to increase the oxygen concentration to enhance the SiC etch rate and depress the fluorine concentration to reduce the Si etch rate. These conditions can be satisfied by choosing low-fluorinated gases (such as CHF_3, $CBrF_3$) and operating at a high oxygen percentage. The latter condition has a secondary advantage of resulting in the chemisorption [10] of oxygen on the Si surface, which will reduce the availability of etching sites and thus decrease the Si etch rate. Oxygen chemisorption is not considered to have a strong effect in the case of SiC, since carbon can react directly with oxygen as indicated in reaction (2). The physical etching mechanism in reactive ion etching is the result of ion bombardment caused by the plasma-induced DC bias on the electrode carrying the samples. Strong ion bombardment will cause a more directional etching and will generate more etching sites by removing carbon and other inhibitors from the surface. However, ion bombardment is less effective at removing material from the sidewall (parallel to the ion direction). Therefore, a carbon-rich surface remains on the SiC sidewall reducing the lateral etching and thus resulting in a more highly anisotropic etching.

For the case of reactive ion etching in SF_6/O_2 plasmas, the largest generation of fluorine is observed compared to the other gases mixtures investigated. This results in the highest (by a factor 2-3) Si etch rate levels and in an etch rate dependence on oxygen percentage in the plasma closely following the fluorine intensity. This confirms previous reports [11] that the main etching mechanism of Si in SF_6 plasma involves the chemical reaction between silicon and fluorine with the formation of volatile compounds. The abundance of fluorine in SF_6/O_2 plasma also affects the SiC etch rate, but to a much lesser degree, by enhancing the removal rate of the carbon-rich layer through reaction (3) shown above. The highest SiC etch rate measured, 53nm/min, occurs for 35%O_2 in SF_6 which coincides with the maximum [F] density. However, the effect of fluorine in the SiC etching process is rather weak as can be seen from the maximum-to-minimum etch rate ratio of only 1.4:1. The SiC-to-Si etch rate ratio in SF_6/O_2 plasma is much smaller than unity for the entire range of parameters investigated. A fairly high (10 to 1) etching anisotropy was observed for SF_6/O_2 etched SiC films, as shown in Fig. 5.

In $CBrF_3/O_2$ plasmas, as shown in Fig. 3, the [F], and to a lesser degree the [O], intensity is very much smaller than in the corresponding SF_6/O_2 plasma. Consequently, the Si etch rate is dramatically reduced, by a factor of approximately 10 to 30. The SiC etch rate was also reduced, but only by a factor of 1.5-2. Therefore, for oxygen percentage in the plasma greater than 60%, the relative abundance of oxygen compared to the near absence of fluorine results in a SiC-to-Si etch rate ratio greater than unity. At $CBrF_3$/75%O_2, a selectivity of 2:1 is observed. Under these conditions, a SiC line edge anisotropy of 6.3:1 was measured (see Fig. 6). The effects of bromine in the etching process still need to be investigated.

In CHF_3/O_2 plasma, the presence of hydrogen depresses mainly the [F] intensity, with the oxygen intensity reaching levels similar to the SF_6/O_2 plasma. The peaks in [F] and [O] intensity occur for 70% and 90% O_2 mixtures, respectively. The Si etch rate variation with O_2% tracked the [F] intensity fairly well. The maximum SiC etch rate, 42nm/min, occurred at 80% O_2 mixture, while the maximum SiC:Si selectivity of 2 was measured at 90%O_2 mixture. The high and almost constant (with O_2%) DC bias found in this case produced the highest anisotropic ratio (12.5:1) achieved.

For selective etching of SiC to SiO_2, the maximum ratio of 1.3:1 was obtained in CHF_3/80%O_2. The optimum reverse selectivity of 1:3.6 was measured in pure CHF_3 plasma. In SF_6/O_2 and $CBrF_3$ plasmas, the SiO_2 etch rate is quite similar to that of SiC and no selectivity was observed.

5. Summary and Conclusions

In summary, we have investigated the effect of various fluorinated-oxygen gas mixtures on the reactive ion etching of SiC and the SiC-to-Si etching selectivity. We have found that in order to obtain selective process one needs to increase the oxygen concentration to enhance the SiC etch rate and to depress the fluorine concentration to reduce the Si etch rate. Optimum selectivities have obtained with CHF_3/90%O_2 and $CBrF_3$/75%O_2 plasma. A summary of optimum selectivity and anisotropy etching conditions is shown in Table I.

Table I. SiC Reactive Ion Etching (RIE)

Gases	Power (W)	Pressure (mTorr)	Anisotropic Ratio *	Selectivity SiC : Si	Etch Rate SiC (nm/min)
CHF_3 +90% O_2	200	20	12.5 : 1	2.0 : 1	41.7
	200	74	5.5 : 1	1.9 : 1	53.3
	300	20	17.0 : 1	2.0 : 1	75
$CBrF_3$+75% O_2	200	20	6.3 : 1	2.0 : 1	37.5
	200	72	7.6 : 1	1.8 : 1	39
	300	20	4.0 : 1	1.3 : 1	50
SF_6 +35% O_2	200	20	10.3 : 1	1 : 26	53.3
	200	200	——	1 : 267	20
SF_6 +90% O_2	200	20	11.0 : 1	1 : 1.8	40

Gases	Selectivity SiC : SiO_2
CHF_3	1 : 3.6
CHF_3+80% O_2	1.3 : 1

* : Ratio of Vertical to Lateral etching

Acknowledgment

The authors would like to acknowledge partial support for this work from the Office of Naval Research, under ONR contract No.N00014-81-K-0605.

References

1. J. Sugiura, W-J. Lu, K.C. Cadien and A.J. Steckl, J. Vac. Sci.,Tech., Vol. B4, 349 (Jan./Feb. 1986).
2. W-S. Pan and A.J. Steckl, Mat. Res. Soc. Symp. Proc., Vol.76, 157(1987).
3. S. Krongelb, IBM, Technical Disclosure Bulletin, Vol.23, No. 2, 828(July 1980).
4. John W. Palmour, R.F. Davis, P. Astell-Burt and P. Blackborow, Mat. Res. Soc. Symp. Proc., Vol.76, 185(1987)
5. S. Dohmae, K. Shibahara, S. Nishino and H. Matsunami, Jap. J. Appl. Phys., Vol.24, No.11, L873(Nov. 1985).
6. J.W. Plamour, R.F. Davis , T.M. Wallett and K.B. Bhasin, J. Vac. Sci. Technol. A, Vol.4, No.3, 590(May/Jun 1986)
7. G. Kelner, S. C. Binari, and P.H. Klein, J. Electrochem. Soc., Vol.134, No.1, 253(Jan. 1987)
8. R. d'Agostino, F. Cramarossa, and S. De Benedictis, Plasma Chemistry and Plasma Processing, Vol. 2, No 3, 213(1982)
9. W-S. Pan and A.J.Steckl, to be published.
10. C. J. Mogab, A.C. Adams, and D.L. Flamm, J. Appl. Phys., 49(7), 3796(1978)
11. A. Picard, G. Turban and B. Grolleau, J. Phys. D:Appl. Phys. , 19, 911(1985)

Index of Contributors

Springer Proceedings in Physics

Managing Editor: H. K. V. Lotsch

MIX
Papier aus verantwortungsvollen Quellen
Paper from responsible sources
FSC® C105338

If you have any concerns about our products,
you can contact us on
ProductSafety@springernature.com

In case Publisher is established outside the EU,
the EU authorized representative is:
Springer Nature Customer Service Center GmbH
Europaplatz 3, 69115 Heidelberg, Germany

Printed by Libri Plureos GmbH
in Hamburg, Germany